Rules
of Thumb
for Home Building,
Improvement, and Repair

Other Books by Katie and Gene Hamilton

Do It Yourself or Not?

Practical Boating Skills

Hand Tool Companion

Build It Together

Gardening Made Easy

Keep It Working

Don't Move, Improve

Fix It Fast, Fix It Right

How to Be Your Own Contractor

Quick-Fix Home Repair Handbook

Wood Projects for the Garden

Wooden Toys

Rules
of Thumb
for Home Building, Improvement, and Repair

Katie and Gene Hamilton

John Wiley & Sons, Inc.
New York • Chichester • Weinheim • Brisbane • Singapore • Toronto

Copyright 1997 by Katie and Gene Hamilton.

Published by John Wiley & Sons, Inc.
Design and production by Navta Associates, Inc.

Library of Congress Cataloging-in-Publication Data

Hamilton, Katie.
    Rules of thumb for home building, improvement, and repair/Katie and Gene Hamilton.
      p. cm.
    Includes index.
    ISBN 0-471-30984-2 (cloth). —ISBN 0-471-30983-4 (paper)
    1. Dwellings—Remodeling—Amateurs' manuals.   2. Dwellings—Maintenance and repair—Amateurs' manuals.   I. Hamilton, Gene. II. Title.
    TH4815.H26   1997
    643'.7—dc21

                                 96-46906
                                      CIP

Printed in the United States of America

10 9 8 7 6 5 4 3 2 1

We'd like to dedicate this book to all the people who help homeowners and do-it-your-selfers wade through the maze of home repair and building products. They're the folks behind the counter at local hardware stores and in the aisles of home centers and lumber yards ready to answer questions and help their customers. Without their depth of knowledge and willingness to share what they know, we'd all be at a loss when it comes to knowing what to buy and how to use it.

Acknowledgments

We want to thank Gary Branson for his help in producing this book. Gary's expertise and insight have been a valuable contribution. And we're ever grateful to our agent, Jane Jordan Browne, who has always supported our efforts and keeps us working.

Contents

Introduction

This is a reference guide that gives you instant access to the information you need when you're involved in a house-building, repair, or improvement project. It's a supplement to other home improvement books because it has inside information that goes beyond the basics. Whether you're building an addition, upgrading a kitchen, or making a small repair, you'll easily find the information you want to know.

We've compiled the material in a user-friendly format so you can find answers and solutions for jobs you're planning.

- Charts give you a reference point to know the sizes, shapes, and features of building materials and components.

- Tables help take the guesswork out of choosing hardware, fasteners, and other home improvement products.

- You'll learn what kind of paint to buy for a specific job, what type of hardware works best with different materials, and what size joists are needed for an addition.

- You'll find formulas to help you determine how many shingles to buy for a roof, what style of molding is appropriate for a job, and the proper joist strength of floor framing.

Each chapter is filled with advice and guidelines to take you through all the projects you might encounter making your house a better place to live—all this, plus tried-and-true procedures that guarantee good results.

Rules
of Thumb

for Home Building,
Improvement, and Repair

1 Wood Building Materials

Lumber is an expensive building material, and choosing the right type for a job can be confusing. For example, a piece of wood is thought of as lumber if it is 2 to 5 inches thick, but it's considered a board if it's less than 2 inches thick. We use the term *lumber* to refer to all types of milled wood.

You'll find many different grades and species of wood, and various manufactured sheet materials made from wood and composites. Here is a rundown of the properties of the most popular types of lumber, plywood, and manufactured wood products you will find in your local lumberyard or home center.

Lumber

Categories of Lumber

Wood is divided into two major categories: softwood, cut from trees that bear needles and produce cones, and hardwood, harvested from trees that have leaves. These types of wood don't necessarily have much relation to the actual hardness of the wood, because some species of softwood, like southern pine, are harder than some hardwoods, like aspen. Generally, softwoods are milled into boards, dimension lumber, and timbers.

Boards. Boards are thin pieces of wood usually 1½ inches or less in thickness and up to 16 inches wide. All boards are sized by their nominal (named) sizes, before they are run through a surface planer and made smooth at the mill.

Dimension lumber. This is a thicker piece of wood, 2 to 4 inches thick and up to 16 inches wide, although few lumberyards stock dimension lumber over 12 inches wide.

Timbers. Timbers are large pieces of wood at least 5 inches thick on each side. Timbers are used for beams, stringers, and posts, as well as in the heavy construction industry and mining operations.

How Lumber Is Measured

Boards, dimension lumber, and timbers are all referred to by their nominal sizes, which are in inches. For example, a 2 × 4 is actually only 1½ inches thick and 3½ inches wide. Table 1.1 lists the nominal sizes and actual sizes of lumber commonly found in lumberyards and home centers.

How Lumber Is Priced

Lumber is commonly priced by the board foot. A board foot of lumber is simply a unit of measurement: 1 inch thick by 12 inches square. For example, 1 running foot of 1 × 12 lumber is equal to 1 board foot; 1 running foot of 2 × 12 lumber is equal to 2 board feet. The nominal sizes of the board are used in all calculations of board feet. Table 1.1 also lists the board feet contained in each running foot of each size of lumber. To calculate the number of board feet, multiply the number in column 3 by the length of the board.

Table 1.2 lists the actual board feet of lumber contained in a given length for commonly used sizes. Look down the left column to find the size of board, then look along the row for the column that has the board's length. The number at the intersection of the row and column is the number of board feet in that size and length of board.

Grades of Lumber

Boards, dimension lumber, and timbers are graded according to the American Lumber Standard. Lumber is graded at the mill, and each area of the country has its own grading associations with different formats for their grade stamps. Even though the stamps may look different, they all contain basically the same information. (See "Deciphering the Grade Stamps on Lumber," on page 4.)

Basically, the grading standards group wood used in housing construction into two categories: select and common. Select grades in most species are the best quality and clear (knot-free). The material is used for cabinetry work and interior trim. Common grades of lumber vary from No. 1, the best-quality lumber with small, tight knots, to No. 4 or 5 (sometimes called economy or utility grade), which can have just about any defect possible.

Common Defects

Knots, warping, checks and splits, and wane are typical defects you'll find, depending on the grade of wood (see figure 1.1).

Knots. Probably the most common defect, a knot is formed when a tree grows around its branches as it matures. Except for the select grades, all grades allow some knots in the wood. The lower the lumber grade, the more knots you can expect to find. Small, tight knots do not affect the strength of the lumber, but large knots or knots close to the

Table 1.1 Lumber Sizes		
Nominal Size	Actual Size	Board Feet per Running Foot
BOARDS		
1 × 2	¾" × 1½"	0.16
1 × 3	¾" × 2½"	0.25
1 × 4	¾" × 3½"	0.33
1 × 6	¾" × 5½"	0.5
1 × 8	¾" × 7¼"	0.66
1 × 10	¾" × 9¼"	0.83
1 × 12	¾" × 11¼"	1.0
DIMENSION LUMBER		
2 × 2	1½" × 1½"	0.32
2 × 3	1½" × 2½"	0.5
2 × 4	1½" × 3½"	0.66
2 × 6	1½" × 5½"	1.0
2 × 8	1½" × 7¼"	1.32
2 × 10	1½" × 9¼"	1.66
2 × 12	1½" × 11¼"	2.0

Table 1.2 Board Feet per Running Foot

Size	Length (in feet)							
	6	8	10	12	14	16	18	20
BOARDS								
1 × 2	1.0	1.3	1.6	1.9	2.2	2.6	2.9	3.2
1 × 3	1.5	2.0	2.5	3.0	3.5	4.0	4.5	5.0
1 × 4	1.98	2.6	3.3	4.0	4.6	5.3	5.9	6.6
1 × 6	3.0	4.0	5.0	6.0	7.0	8.0	9.0	10.0
1 × 8	4.0	5.3	6.6	7.9	9.2	10.6	11.9	13.2
1 × 10	5.0	6.6	8.3	10.0	11.6	13.3	14.9	16.6
1 × 12	6.0	8.0	10.0	12.0	14.0	16.0	18.0	20.0
DIMENSION LUMBER								
2 × 2	1.9	2.6	3.2	3.8	4.5	5.1	5.8	6.4
2 × 3	3.0	4.0	5.0	6.0	7.0	8.0	9.0	10.0
2 × 4	4.0	5.3	6.6	7.9	9.2	10.6	11.9	13.2
2 × 6	6.0	8.0	10.0	12.0	14.0	16.0	18.0	20.0
2 × 8	7.9	10.6	13.2	15.8	18.5	21.1	23.8	26.4
2 × 10	10.0	13.3	16.6	19.9	23.2	26.6	29.9	33.2
2 × 12	12.0	16.0	20.0	24.0	28.0	32.0	36.0	40.0

edge of the board do. Loose knots that are formed when the tree grows around a dead branch eventually fall out. All knots require sealing prior to finishing, because their high pitch content will eventually bleed through most paints or stains.

Warping. Warping is caused by uneven drying of wood. This condition is due to the grain pattern, which depends on the part of the tree the log is cut from. Lumber with a vertical grain pattern at the end of the board tends to shrink uniformly on all sides as it dries. Lumber with a horizontal grain pattern tends to warp, twist, and bow. When purchasing wood, look at the end of the board and choose lumber that has a vertical grain pattern.

Figure 1-1 Common Defects

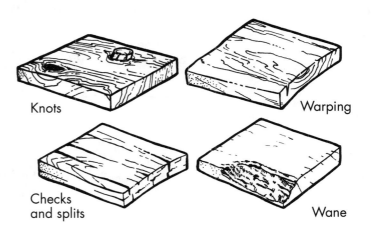

Knots

Warping

Checks and splits

Wane

Checks and splits. A check or split is a crack in the lumber running with the grain, which often occurs when wood dries. It usually appears near the end of the board and often doesn't form until after the lumber is graded and shipped to the lumberyard. Unless the board is being used full length, small splits will not affect its use.

Wane. Wane is an uneven, rough edge on boards cut from the outer, bark part of the log. The wane occurs when the bark falls off. This defect is not found in better grades of wood, but it is found in economy or utility grades.

Deciphering the Grade Stamps on Lumber

Lumber is stamped with the mill number, lumber grade, moisture content, name of inspection service, and, sometimes, the wood species (see figure 1.2). Moisture content is an important factor. After a tree is cut, the wood gradually loses most of its moisture, causing the wood's cells to shrink. This overall drying process causes wood to change dimensions. Drying out can take many months if the lumber is air-dried. Today most lumber is dried in kilns, large ovens that control heat and humidity and dry wood quickly.

Most lumber today is kiln-dried to a moisture content of less than 19 percent and is marked S-DRY (surfaced dry). Lumber marked S-GRN has a moisture content of over 19 percent and is considered green wood. Lumber stamped MC 15 has a moisture content of not more than 15 percent.

All the grade stamp can tell you about the moisture content of the wood is the content when the lumber left the mill. How and where lumber is stored has a lot to do with its moisture content. Lumber stored outdoors, exposed to the elements, may have a higher moisture content than the grade

Figure 1-2 Typical Grade Stamps for Lumber

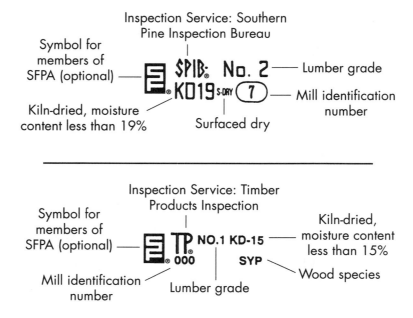

Working with a Building Inspector

A building inspector relies on grade stamps to determine the quality of framing lumber. Make sure that the grade stamp is visible on most of the framing members. While knots and other imperfections are allowable in the grade, do not let knotholes or cutouts in the framing member accommodate plumbing, electrical, or heating ducts. Such use may cause an inspector to question the integrity of the structure.

stamp indicates. A pile of lumber stacked next to a building in the shade below a leaky gutter will certainly have its moisture content affected.

Dimension Lumber

Most house framing and construction use dimension lumber, available in most parts of the country in three basic classes.

Light framing lumber. Light framing lumber, which is less than 4 inches in height and width, is generally available in construction, standard, utility, and economy grades. In many areas the top two grades are grouped together in a single standard or better grade, which is for house wall-framing jobs. Avoid the economy-grade material, which usually has twists and bows making it unusable for general framing.

Studs. Studs are at least 2 inches thick and up to 6 inches wide, and are used for supporting elements of a wall or partition, including load-bearing walls. It makes up the framework that wallboard, paneling, and the like are attached to. Stud is also a grade of lumber signifying that it can be used as a structural part of a building.

Structural light framing lumber. Structural light framing lumber is used for joists, rafters, and headers. This material is available in sizes from 2 to 4 inches thick and wide. The best grade is select, followed by No. 1, No. 2, No. 3, and economy. Most building codes require that No. 2 or better is used.

Molding

Molding is the decorative trim traditionally used to hide joints between walls and ceilings, around windows and doors, between floors and staircases, and in other areas where two surfaces meet. Most moldings are expensive because they're milled from clear select wood, the best-quality wood available.

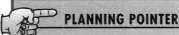

Choosing Grades of Lumber

In most situations it doesn't pay to use a better grade of lumber than necessary. Although the middle grades have cosmetic defects, they are just as strong as the top grades and are less expensive. Economy grades of lumber usually push the lower limit of the grade level and don't represent value, since much of the lumber ends up unusable (or usable only as firewood).

Interior Molding

While lumberyards carry a selection of the most popular styles of interior molding, often you'll have to order the more intricate patterns (see figure 1.3), especially if you need long lengths. Suppliers of specialty architectural wood often offer a selection of molding patterns and wood species, including casings with plinth blocks (window and door moldings with decorative corner blocks) for replacing or adding traditional molding in older homes.

Sometimes you'll find common casings and base moldings (used between floors and walls) in finger-jointed pieces. These are less expensive since they're made from shorter pieces of wood joined together with glued, interlocking joints. Because they're made of mismatched wood, finger-jointed moldings are meant to be painted, not stained.

Moldings are sized according to their width. For example, 3½-inch base molding is 3½ inches wide and 9/16 inch thick, and is usually available in stock lengths up to 16 feet. Wide ornamental moldings, such as crown molding (used between ceilings and walls), are dimensioned by the widest part. However, since it is installed at an angle, the molding's actual height on the wall will be somewhat less.

Ceiling moldings.   Ceiling moldings include crown, bed, and larger cove moldings. These are applied at a 45-degree angle between the ceiling and wall. This allows the molding to be milled from a smaller piece of lumber, but just as important, the molding can be fitted to uneven surfaces. Stock ceiling moldings up to 5 inches wide can be found. The thinner pieces of molding milled from square stock (4 × 4s and the like) will bend to fit and conceal imper-

Figure 1-3   Patterns of Interior Molding

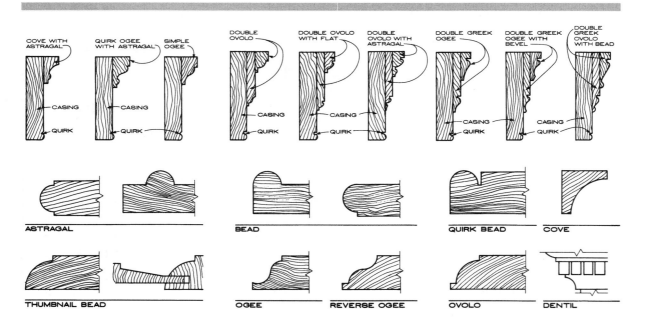

fections in the walls and ceilings. For a custom pattern or to match an existing one, you can build up several pieces of molding by gluing them together before installation.

Base moldings.   Base moldings, or baseboards, are installed on the wall at the floor to protect the wall and hide the joint between the wallboard or plaster and the floor. Traditionally, base molding is made up of three parts: the base, a broad, flat board; the base cap, which sits on top of the base; and the base shoe, applied along the floor. The base cap and base shoe are smaller, more flexible moldings that are not only attractive but cover the gap between the stiff base and an irregular wall or floor.

Casings.   Casings are wide, thin moldings installed around the edges of windows or doorjambs. Like a picture frame, casings hide the gap between the framing of the window or door and the wallboard. Stops are thin strips of molding installed around the perimeter to hold a window in its frame and to prevent a door from opening in the wrong direction in the jamb. Stools, or windowsills, are members used to trim out traditional windows.

Wall moldings.   Wall moldings are not only decorative but also protect the wall and conceal joints. Chair rails protect the wall from being hit by the back of a chair. Ply caps cover raw edges of plywood paneling and hide the joint between the top of the panel and the wall, as in wainscoting. Panel molding, panel strips, and mullion casings are used to hide joints in paneling and are applied to walls for decoration.

Astragal molding is designed to cover joints between paneling, double-hung windows, and doors. Quarter round hides corner joints in paneling, and half round is often added between paneling to conceal joints or to other surfaces for decoration.

These patterns are all stock items available at most lumberyards.

Handrails.   Stock handrails are available in many patterns and are used in stairways, on walls in hallways, or wherever there is a need for a molded wood shaped to fit the hand.

Exterior Molding

Brick molding and drip caps.   These moldings are used to trim out the exterior of a window or doorjamb. Brick molding is designed in a shape that hides the joint between the window frame or doorjamb and masonry. Drip caps are installed over brick molding or other exterior casing to direct water away from the window or door.

Screen molding.   Screen molding is used in wooden door and window screens to anchor the screening fabric and conceal rabbets, cuts made in the frame where the screening is nailed or stapled in place.

Glass bead.   Glass bead is used instead of putty to hold window glass in place. It's a decorative molding that's used in many historic renovations.

Lattice.   Lattice is a frequently used material. As a thin wood cut in small dimensions, it's useful for many applications, including building trellises. It ranges in thickness from $\frac{1}{2}$ to $1\frac{1}{16}$ inches and in widths from $\frac{3}{4}$ to $2\frac{3}{4}$ inches, so it is often used to build up a crown molding or fill gaps where other molding is too large.

Metal-clad wood-core molding.   Commercial builders have been using metal-clad accent moldings in hotels, office buildings, and restaurants for years to create a striking contrast. Consumers can get the material cut to order in brass, chrome, copper, or bronze. The metal cladding is polished and sealed with a transparent lacquer coating, which protects it from fading, oxidizing,

RULE OF THUMB

Cutting Metal-Clad Molding

To get a clean cut at the edges of metal-clad molding, use a very thin saw blade with 90 to 100 teeth.

and tarnishing. Metal-clad molding can be used alone or in combination with other patterns to create custom architectural decorations on walls and ceilings, around doors and windows, or anywhere molding is used.

All finishes are protected with peelable plastic film so they're not damaged during handling and installation. Lift the edges of the plastic film before installing the molding so you can easily remove the film. Like traditional wood molding, metal-clad molding is cut with a table saw. But it's installed with a quick-drying contact cement, not with nails or screws.

Hardwood

Hardwood is most often used to craft cabinets, bookcases, furniture, and moldings and other decorative trim. Most lumberyards and home centers carry a limited supply of hardwood (see table 1.3). Most areas have a lumberyard or other source that specializes in stocking hardwood. Hardwood is also available from mail-order catalogs.

Lumberyards that specialize in hardwood usually carry random lengths and widths. Since hardwood is rarely used to build structures, it is graded for appearance, not strength. Factory lumber, the type used in the furniture industry, is seldom found in lumberyards, but the dimension and finish grades are.

Moldings and turnings like balusters are made from finish-grade hardwood. Dimension grades are available in most lumberyards. Dimension-grade hardwood is broken into five basic grades: firsts, firsts and seconds, seconds, select, and common. Very few suppliers handle firsts, which must be at least 4 inches wide and 7 feet long and contain not more than $8\frac{1}{3}$ percent waste per board. Most lumberyards do carry the next grade, firsts and seconds (FAS). These are wide, clear boards, at least 6 inches wide and 8 feet long. The next grade, select, is at least 4 inches wide and 6 feet long; common grade is narrower and shorter.

Home centers stock standard precut sizes of hardwood in limited quantities and species. Red oak, poplar, maple, ash, walnut, and cherry are some of the available woods. These boards are clear and finished on all sides and are available in thicknesses from $\frac{1}{4}$ to $\frac{3}{4}$ inch and in widths up to 12 inches. Several national lumber producers have hardwood programs that include display racks stocked with the most popular species and sizes. In addition to a selection of hardwood boards, these display racks sometimes contain a selection of hardwood boards edge-glued together to form assemblies that can be used for tabletops or countertops.

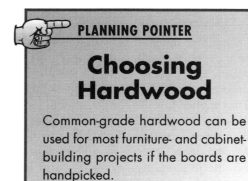

PLANNING POINTER

Choosing Hardwood

Common-grade hardwood can be used for most furniture- and cabinet-building projects if the boards are handpicked.

Table 1-3 Readily Available Hardwood Carried by Home Centers and Lumberyards

| Species | Board Sizes | | |
	WIDTH	LENGTH	THICKNESS
Red oak	2" to 12"	2' to 8'	¼", ½", ¾"
Poplar	2" to 12"	2' to 8'	¼", ½", ¾"
Maple	2" to 12"	2' to 8'	¼", ½", ¾"
Ash	2" to 12"	3' to 8'	¼", ½", ¾"
Walnut	2" to 12"	3' to 6'	¼", ½", ¾"
Cherry	2" to 12"	3' to 6'	¼", ½", ¾"

Redwood and Cedar

Redwood and cedar both have natural qualities that make them excellent choices for interior or exterior applications. These softwoods are naturally resistant to rot and decay. In both species the most durable wood is the heartwood, cut from the center of the tree. When shopping for wood that will be in continuous ground contact, such as retaining walls or fence or deck posts, specify heartwood.

Both redwood and cedar are easy to work with, resist splitting when nailed, hold paint or stain well, and are the premium wood choices for resistance to shrinkage and warping. Tests by the U.S. Forest Products Association show that redwood has the least shrinkage of any commercial American wood. Either wood is a superior choice for exterior projects, such as decks, siding, or fences. Left unpainted, cedar and redwood will gray to an attractive natural finish. Both are low-maintenance materials because they hold finishes well.

Both redwood and cedar are available in all the structural lumber sizes, including 1×, 2×, and 4× thicknesses. They are also available in a wide variety of siding styles.

Grades of Redwood

Redwood is available in more than 30 grades. The important points for the consumer to remember are that the architectural grades are the most resistant to decay and insect attack. The architectural grades are clear all heart, clear, and B grade. These grades should be used for any project that will be in continuous ground contact, and for siding, trim, paneling, or cabinetry.

The more economical grades of redwood are called garden grades. These are construction heart, construction common, merchantable heart, and merchantable. Use these grades where knots or color variations are not important. Typical uses for garden grades are decks, fences, and other garden projects.

For the projects where there will be no ground contact, choose the least expensive garden grades. These grades are less desirable because they contain some sapwood, the newest growth closest to the outside of the tree.

Treated Wood

As environmental concerns have grown, harvest of such naturally durable softwood species as redwood and cedar has been drastically curtailed. But by chemically treating more common, fast-growing species of softwood, such as pine and fir, these lumber species can be made more resistant to decay and can be substituted for the naturally durable species.

Wood is treated when it is injected with chemicals, such as chromated copper arsenate (CCA), to make it more resistant to rot and decay caused by weather, moisture, and

RULE OF THUMB

Using Treated Wood

Because treated wood may contain high amounts of liquid chemicals when it reaches the consumer, further drying is often necessary before building with it. Stack the lumber in layers separated by wood lath so that air can circulate through the pile and dry the lumber.

insects. Because of its durability, treated wood is used commercially for docking and seawalls in both fresh- and saltwater applications, and has been approved by the Environmental Protection Agency (EPA) for use in public parks and playgrounds.

PLANNING POINTER

Choosing the Right Grade of Lumber

The amount of chemicals injected into the wood varies depending on the intended use, and the durability of the wood depends on the amount of chemicals it contains. Look for grade stamps that indicate whether the wood is intended for continuous ground contact, such as in-ground deck posts or retaining walls, or for above-ground projects, such as decking or fencing. These stamps vary but should indicate, in addition to exposure conditions, the grade of lumber, inspection service, treatment company and location, year of treatment, chemical used for treatment, and chemical retention level (see figure 1.4).

Figure 1-4 Typical Grade Stamp for Treated Wood

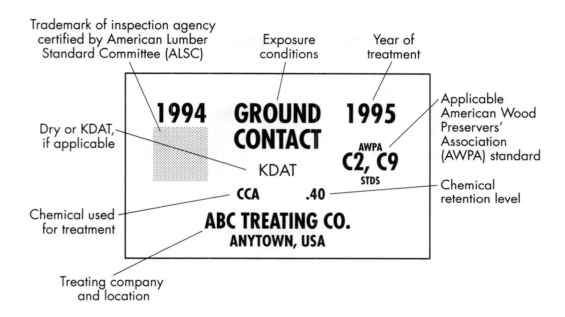

For consumer applications, treated wood is used to build foundations; children's swings, sandboxes, and play sets; and decks, retaining walls, and other outdoor structures.

Treated wood is available in all the same stock sizes as untreated wood, that is, in 1×, 2×, and 4× thicknesses and common widths, and in plywood in 4 × 8 sheets, either ½ inch or ¾ inch thick. Special treated-wood products, such as lattice, turned posts, railings, balusters, and finials, are manufactured for outdoor building projects.

RULE OF THUMB

Sealing Treated Lumber

Treated lumber is often manufactured using species of lumber having an open wood grain. This grain makes the wood subject to cracking and splitting if it is not sealed with a clear water repellent. Some products are sealed at the factory and do not need an immediate seal coat. Other manufacturers advise letting the wood season for a few months before applying the water repellent. To avoid cracking and splitting, all treated wood products require periodic application of a water repellent, typically in the spring and fall, as preventive maintenance.

Safety Tips for Working with Treated Wood

Treated lumber has been approved by the EPA for any ordinary household application, including use as garden stakes or borders, and for playground applications at schools and parks. However, as with any chemical product, reasonable care should be taken when building with treated wood.

● Do not use treated wood for countertops or any food-preparation surface. The U.S. Food and Drug Administration does not recommend using *any* wood as a food-cutting surface because of the danger of getting contaminated food particles into the wood.

● When working with treated wood, wear a dust mask to avoid inhaling sawdust, and wear goggles to protect your eyes from sawdust and chips.

● Wash your hands after handling treated wood and before eating, drinking, or smoking.

● Treated lumber does not give off any fumes when used indoors, making it usable as a mudsill, for example, for extra protection against wood rot.

● Do not burn treated lumber because the ash and the smoke may be toxic. Dispose of it in an approved landfill.

Plywood

Plywood is a composite of veneers glued together with the grain of each layer laid at right angles to each other to increase strength. Plywood is less expensive than solid wood and preferred by many carpenters because of its price and lighter weight.

All plywood is not equal, which accounts for its diversity of uses. At one end of the spectrum there is everyday construction-grade plywood used as floor underlayment to cover subfloor irregularities and as roof and siding sheathing—the first layer of exterior wallcovering. At the high end you'll find beautiful hardwood-veneer plywood used by woodworkers to make furniture and cabinetry. In between there are different grades and various types of the material, such as plywood for exterior use or marine use.

Grades of Plywood

Since high-grade plywood is expensive and really not necessary for most uses, home centers and lumberyards usually stock sheets with a combination of veneer grades.

These veneer grades refer to the quality of the wood used in the face plies (outer veneers). The best face-ply rating is A, meaning the face ply is free of open knots or other blemishes. Defects are neatly repaired, and the surface is smooth and paintable. Face-ply ratings of B and C allow some blemishes, such as tight knots and minor splits, in the face plies. Face plies so rough that open knotholes are permitted are rated D. For sheathing plywood, where appearance of the plywood is not important, the face-ply rating is C-D, meaning that one side of the plywood is C grade and the other, D. Plywood panels rated grade A-C or A-D are blemish-free on one side and less perfect on the other, so they're suitable for most projects.

In addition to letter grades, plywood is also graded by expected use. Sheathing panels are of course used for sheathing on houses, and Sturd-I-Floor panels are used for flooring.

Plywood panels are graded not only for appearance but also for weather exposure. Exterior panels have waterproof glue holding the veneer laminates together and are designed to withstand permanent exposure to the weather. Panels rated as exposure 1 are made with waterproof glues but are designed for limited exposure to the weather. Exposure 2 panels are designed for limited exposure to the weather and humidity. Interior panels are not manufactured with waterproof glue and should be used only for interior applications.

How to Read Grade Stamps for Plywood

Plywood manufactured by members of the American Plywood Association (APA) carries a grade stamp that tells you all you need to know to make your selection. But to make your selection, you must know how to read the grade stamp.

First, the mark APA in figure 1.5 identifies the manufacturer as a member of the APA. The panel-grade mark identifies the use of the plywood, in this case, sheathing plywood, or fir plywood intended for use as floor, wall, or roof sheathing. The large numbers show the allowable span (maximum distance between rafters or joists) for application on roofs or floors. A span rating of 32/16 indicates a 32-inch span for use on roofs, but only a 16-inch span for use on floors. Next to the span rating is the thickness of the plywood, in this case, $^{15}\!/_{32}$ inch.

The next mark shown in figure 1.5 is the exposure rating. This refers to the type of glue used to bond the plies and its resistance to water or moisture. Plywood rated as exposure 1, as in the figure, should be used only where it will later be covered by siding or shingles. Below the exposure rating are the mill number and product number and other APA codes for quality assurance and performance rating.

Figure 1-5 Typical Grade Stamp for Plywood

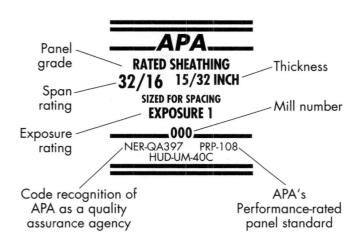

Panel grade
Span rating
Exposure rating

APA
RATED SHEATHING
32/16 15/32 INCH
SIZED FOR SPACING
EXPOSURE 1
000
NER-QA397   PRP-108
HUD-UM-40C

Thickness
Mill number

Code recognition of APA as a quality assurance agency

APA's Performance-rated panel standard

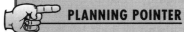 **PLANNING POINTER**

Choosing Plywood for Furniture and Cabinets

Ordinary finish plywood may have voids or knot-holes in the core (center) plies. Center-ply voids will leave ragged edges called "core gaps" when cut. Buy solid-core plywood, called "lumbercore," for furniture or cabinet projects.

Finish Plywood

Finish plywood is intended for interior use in cabinets, furniture, or the like, and thus carries no span rating or exposure rating as sheathing plywood does. Finish plywood is sanded on both sides. The best grade is A-A, meaning both sides are unblemished and can be used where both sides will be finished and exposed. For example, A-A

plywood is used for cabinet doors, where the inside face will be exposed when the cabinet doors are open. For furniture, where only one side will be exposed to view, A-B or A-C plywood may be used.

Manufactured Wood Products

As the name implies, manufactured wood products are made from wood by-products, such as wood chips, bonded together with resins or other adhesives. This group of wood products, called composites, includes particleboard, waferboard, oriented-strand board (OSB), and hardboard. See table 1.4 for the sizes and uses of composite panels made from these materials.

Composite materials are heavy and can absorb moisture, so use them in jobs where their exposure to water won't create a problem. For example, don't use particleboard

Table 1.4 Composite Panels

Material	Available Sizes[a]	Uses
Particleboard	Length: 8', 10', 12' Thickness: 3/8", 1/2", 5/8", 3/4"	Subflooring, wall and roof sheathing, shelving, underlayment
Waferboard	Length: 8', 10', 12', 14', 16' Thickness: 7/16", 1/2", 5/8", 3/4"	Subflooring, wall and roof sheathing, shelving, underlayment
Oriented-strand board (OSB)	Length: 8', 10', 12' Thickness: 3/8", 7/16", 1/2", 9/32", 5/8", 23/32", 3/4"	Subflooring, wall and roof sheathing
Hardboard	Length: 8', 10', 12' Thickness: 1/8", 1/4"	Cabinet backs, sliding doors, signs Also available prefinished and with holes, commonly known as Peg-Board

[a] All panels are 4' wide.

as an underlayment for ceramic tile in a bathroom.

Particleboard

Adhesives are combined with wood particles and chips to make particleboard. It is available in 4-foot-wide sheets from 8 to 12 feet long and from ⅜ to ¾ inch thick. Often called chipboard because of the fragments or chips of wood, particleboard is used as carpet underlayment, base material for cabinet construction, and precut shelving of various widths.

Grades of Particleboard

While there are low, medium, and high ratings for the density of particleboard, most of what's sold in a lumberyard or home center is medium grade. Numbers on the grade stamp identify the bonding resin: 1 is urea formaldehyde, suitable for interior use, and 2 is phenol formaldehyde, for exterior use. Another number on the grade stamp identifies fastener strength, with 3 being the greatest. Thus, a sheet marked 1-M-3 is a medium-strength sheet for interior use (bonded with urea formaldehyde) and with a high ability to hold fasteners.

Waferboard

Random waferlike shapes of wood are mixed with resins to create waferboard. Waferboard comes in 4-foot-wide sheets from ⁷⁄₁₆ to ¾ inch thick and up to 16 feet long. Waferboard is used as underlayment, shelving, and various other exterior applications.

Oriented-Strand Board (OSB)

Oriented-strand board (OSB) is a combination of strands of wood fibers bonded together with phenolic resins. This material is less expensive than comparable sheets of plywood. OSB is available in 4-foot-wide sheets from ⅜ to ¾ inch thick and 8, 10, or 12 feet long. Sheets can have either straight or tongue-and-groove edges. OSB sheets are graded by the American Plywood Association for use in residential construction and for wall and roof sheathing and subflooring.

Hardboard

Hardboard is made from wood pulp that is pressed into thin sheets. Hardboard comes in 4-foot-wide sheets from ⅛ to ¼ inch thick and from 8 to 12 feet long.

Hardboard is available in several forms, but most lumberyards stock the standard grades, which are finished smooth on one side (S1S) and textured on the other. It is also available smooth on both sides (S2S). Standard grades also come with

 **RULE OF THUMB**

Safe Use of Composite Materials

Because the resins and adhesives contained in some composite materials have harmful vapors, work with these materials in a well-ventilated area.

RULE OF THUMB

Cutting Composite Materials

When cutting composite materials, have the saw teeth cut into the finish side for the smoothest, fuzz-free edge. When using a handsaw, place the material faceup; when using a power saw, position it facedown. The resins and adhesives dull cutting tools, so use a carbide saw blade and router bits.

evenly spaced holes drilled through them. This material is commonly known as Peg-Board. A tempered version, impregnated with resins to make the material more water resistant and stronger, is available. Hardboard also comes with painted surfaces, called enameled hardboard, and with embossed patterns.

2 Fasteners and Adhesives

Fasteners

A fastener is anything that joins two surfaces together. Although there are many types of fasteners, the basic ones include nails, screws, nuts and bolts, washers, anchors, metal framing connectors, staples, and rivets. The success of the connection depends on the material of the surfaces and the properties of the fastener.

Nails. Nails are common fasteners driven with a hammer to connect wood members together.

Screws. Screws have greater holding power than nails because they are threaded. Screws join wood to wood or metal to metal, depending on what they are made of. They remove easily because they come out with a turn of the head.

Nuts and bolts. Nuts and bolts are a fastening system used for joining wood and metal. The bolt is a threaded fastener that works with a retaining nut. Often a washer is used to reinforce the nut so it doesn't pull out of the material.

Washers. Washers are metal disks that extend the holding power of fasteners beyond the area covered by the nut or head of a bolt or screw.

Anchors. Anchors come in a variety of shapes and sizes. They're designed to fasten objects to walls, ceilings, or floors.

Metal framing connectors. Metal framing connectors are fasteners that create a strong connection for joists, beams, posts, and other construction assemblies.

Staples. Staples are U-shaped tacklike connectors driven into wood by a staple gun.

Rivets. Rivets are nail-like connectors installed with a rivet tool that presses the rivet through the surface, creating a bond on thin materials like gutters or canvas.

Nails

The simple nail has proved to be the easiest and cheapest way to connect two pieces of wood. There are common nails, also called wire nails, and special-use nails designed for particular applications. The two key features of a nail are its head and shank, which determine its

RULE OF THUMB

Organizing Fasteners

Get organized with a compartmentalized plastic organizer for various sizes of nails, screws, bolts, and so on. Or make your own using jars, small cans, and so forth. As you accumulate fasteners, remove them from the packaging they're sold in and store them so they're easy to locate and access.

use. Small-head nails are used where the work is visible and the head can be concealed with filler. A wide head provides a larger striking surface and is useful for fastening soft material, such as a roof shingle. A nail can be smooth, round, serrated, or annular (ringed). A nail with a broad, flat head and a wide-diameter, serrated shank drives itself deeply into wood fibers and thus has great holding power, while a nail with a smooth, small head can be easily nailed and removed from the wood.

Nails are categorized and sized by their length, with a letter *d* used to designate their size. You'll hear a 3-inch-long nail referred to as a tenpenny nail, which is expressed as 10d. Penny sizes and nail types are shown in figure 2.1.

Types of Nails

Different metals are used to make nails of different types. Those made of steel are for general construction, like joining wood and building materials. Aluminum nails are used for building and for installing aluminum siding and roofing. Hardened steel nails are used for masonry and flooring installations. Copper, bronze, silicon bronze, nickel silver, monel, and stainless steel nails are corrosion resistant, so they're used in the marine industry for building boats.

Nails can be finished in a variety of ways. Bright nails are most common; they're used indoors and where there's no moisture. Galvanized and cadmium-plated nails prevent rust stains. Painted nails match siding or other building materials.

Common nails. They're called common because they're the basic nail used in construction framing and rough carpentry. Common nails have a large nail head and range from 2d (1 inch long) to 60d (6 inches long).

Box nails. These are thinner nails with a smaller diameter than common nails, so they're lightweight and less tiring to drive. Their slightly blunted tips make box nails a good choice for thin wood, which could be split by common nails.

Finishing nails. Fine carpenters use finishing nails for installing woodwork and trim, building cabinets, and especially when connecting thin pieces of wood. The thin nails are made of a lighter-gauge wire ranging from 2d to 16d. Finishing nails are installed with a nail set, which drives the small nail head beneath the surface of the wood.

Finishing nails smaller than 2d are called brads, which measure ½ to 1½ inches long. They can be hammered in place or installed with a brad gun, which shoots the small metal nail into the wood at high speed.

Casing nails. The trim around the outside of doors and windows is called casing, and so are the nails used to install it. Casing nails are a special-use finishing nail with a shank like a box nail and a head like a finishing nail.

Special-Purpose Nails

Special-purpose nails are designed to join a variety of building materials so that they stay fastened, ensuring a secure installation. They're sold in various sizes and in both large and small quantities.

ROUGH CONSTRUCTION

NAME	SHAPE	MATERIAL	FINISH
COMMON	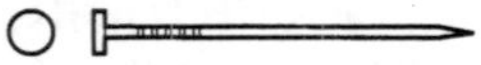	Steel or aluminum	Smooth
ANNULAR		Steel, hardened steel, copper, brass, bronze, silicon bronze, nickel silver, aluminum, monel, or stainless steel	Bright, hardened
HELICAL			
COMMON CUT STRIKE		Steel or iron	Bright or zinc-coated
DOUBLE-HEADED	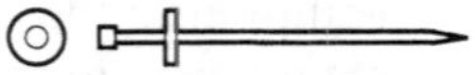	Steel	Bright or zinc-coated
	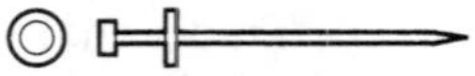	Aluminum	Bright
SQUARE	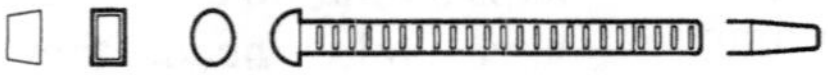	Steel	Smooth, bright, zinc-coated
ROUND WIRE	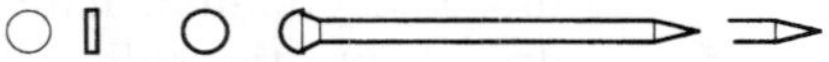		
ANNULAR	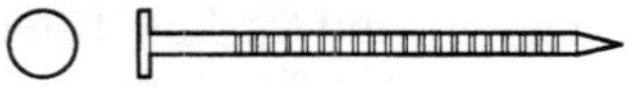	Aluminum	Bright or hard

ROOFING

NAME	SHAPE	MATERIAL	FINISH
SIDING & SHINGLE		Steel, copper, or aluminum	Smooth, bright, zinc- or cement-coated
ROOFING (BARBED)	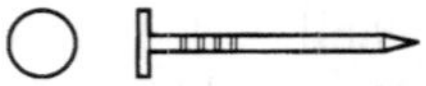	Steel or aluminum	
ROOFING	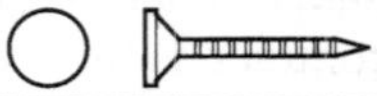	Steel	Bright or zinc-coated
NONLEAKING ROOFING	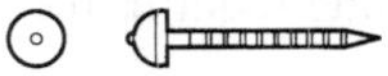		
SHINGLE NAIL	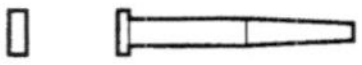	Steel or cut iron	Plain or zinc-coated
CUT SLATING (NONFERROUS)	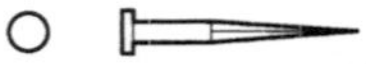	Copper, muntz metal, or zinc	
GUTTER SPIKE (ROUND)		Steel	Bright or zinc-coated
GUTTER SPIKE (ANNULAR)	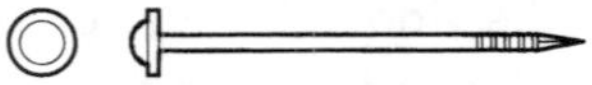	Copper	Bright

FINISH WORK

NAME	SHAPE	MATERIAL	FINISH
WALLBOARD	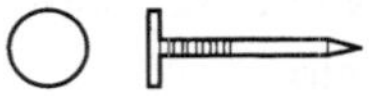	Steel or aluminum	Smooth, bright, blued, or cement-coated
FINE NAIL		Steel	Bright
LATH			Blued or cement-coated
LATH	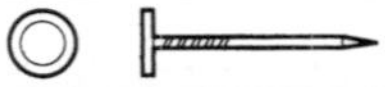	Steel or aluminum	Smooth, bright, blued, or cement-coated
CASING OR BRAD			Bright or cement-coated
FINISHING		Steel	Smooth

Figure 2-1
Penny Sizes and Nail Types

Drywall nails.   Designed specifically to secure gypsum wallboard to framing, these nails have smooth or annular shanks. For most installations they've been replaced by drywall screws, which eliminate the problem of nails pulling away from the wood framing, often called nail popping. This problem is caused by the natural movement and vibrations of a building and the expansion and contraction of framing lumber.

Masonry nails.   To fasten wood into concrete or cement block, a masonry nail is the best choice. The nail is heavy and hardened and has various types of shank, including round, fluted, flat, or square.

Double-headed or duplex nails.   For a temporary construction project, or anything that will be knocked down or dismantled, a duplex nail makes removal easy because it's actually two nails in one. This nail has a flange about ½ inch down the shank from the head. You drive the nail into the wood up to the flange, which acts like a stopgap. The head protrudes, so it's easy to remove the nail with a claw hammer when it's time to dismantle the assembly.

Flooring nails.   To install strip flooring, use either hardened cut nails with no heads or spiral nails. Because cut nails are hardened, they have more holding strength. Spiral nails grip the wood fibers as they are driven into the sides of the floor planks.

Siding nails.   The wide variety of house siding requires an equally diverse type of nail to install it. Siding nails, in 6d and 8d, are resistant to corrosion.

Traditional wood lap siding requires nails similar to a finishing nail. Siding nails have sinker heads that are thinner than those of common nails, so they can be driven into the siding below the surface and the hole filled with wood putty. Annular finishing or siding nails are used on plywood siding. Shingle nails used for roofing hold cedar shingles and shakes in place.

For aluminum siding, an aluminum nail with a large flat head is used so the material hangs loosely and can expand and contract as the weather changes. Vinyl siding requires the same type of nail.

 **RULES OF THUMB**

Driving Nails

To drive finishing nails, first drive the nail into the wood, then use a nail set to drive the nail head beneath the surface of the wood. Put the tip of the nail set on the nail head and strike the nail set with a hammer. One good blow is usually all that's needed.

To secure a masonry nail in brick, first drill the hole with a masonry bit, then stuff the hole with a wad of steel wool to give the nail something to bite into.

To join wood pieces together, nail the thinner piece of wood to the thicker one. Choose a nail that is three times the thickness of the thinner board, because a good two-thirds of the nail will be driven into the thicker piece.

Always wear eye protection (glasses with side panels or safety goggles) when you're driving nails to prevent splinters of wood from injuring your eyes.

Roofing nails. Nails for asphalt and fiberglass shingles are usually 6d and are easily identified by their wide head. They are galvanized and can be barbed, annular, or spiral. Some have washers.

Sheathing or underlayment nails. To prevent nails in floor underlayment or sheathing from pulling out, annular nails with small flat heads are used.

Screws

Screws join wood or any other materials together better than nails because of their strength and holding power. Two pieces of lumber nailed together can pull apart as the wood gains or loses moisture and expands or shrinks. Constant pressure trying to pull the pieces apart can also loosen the nail. Because it's threaded, a screw is more deeply embedded into the fibers of the wood. Thus it's less likely to pull loose and more likely to stay in place. The threads dig into wood or any other material as the screw is driven into a predrilled hole called a pilot hole. Most screws are made of steel and range from ¼ to 5 inches long.

Types of Screws

There are various types of screws determined by the shape of the screw head and the material the screw is designed for. They are driven by a screwdriver designed to fit the screw head. Flat-head screws are slotted and require a straight-slot screwdriver, also called a standard or flat-blade screwdriver. A screw with a crisscross slot requires a Phillips-head screwdriver designed to fit the X in the screw head. There are other specialty screws, like Torx screws used on automobiles and lag screws (also called lag bolts), which are driven with a wrench. Most screws are made of steel, but you'll also find brass, stainless steel, and aluminum screws, which resist corrosion.

Wood screws. Wood screws are specified by their gauge, which is the diameter of the unthreaded shank below the head. Screw gauges range from 0 to 24; length is in inches. Screw heads can be flat, round, or oval, and the groove in the head can be slotted or crisscrossed. (See figure 2.2.)

Use a flat-head screw when the screw head will be flush with or recessed below the surface. Flat-head screws are often countersunk by drilling a shallow hole the same diameter as the screw head so the screw head can be concealed with a

Figure 2-2 Wood Screws

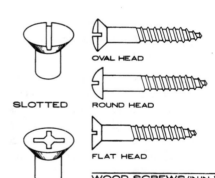

DIA.	DECI. EQUIV.	LENGTH
0	.060	¼ − ⅜
1	.073	¼ − ½
2	.086	¼ − ¾
3	.099	¼ − 1
4	.112	¼ − 1 ½
5	.125	⅜ − 1 ½
6	.138	⅜ − 2 ½
7	.151	⅜ − 2 ½
8	.164	⅜ − 3
9	.177	½ − 3
10	.190	½ − 3 ½
11	.203	⅝ − 3 ½
12	.216	⅝ − 4
14	.242	¾ − 5
16	.268	1 − 5
18	.294	1 ¼ − 5
20	.320	1 ½ − 5
24	.372	3 − 5

WOOD SCREWS (IN IN.)

RULE OF THUMB

Using Wood Screws

For the best results in installing any wood screw, predrill a pilot hole. Make the pilot hole slightly smaller than the screw so the screw bites into the wood fibers.

wooden plug. A round-head screw is usually used with a washer designed for thin wood. The washer protects the wood from being penetrated by the screw head. Oval-head screws are used for decorative purposes, because they protrude above the surface of the wood. Pan-head screws are another type of decorative screw, shaped like an inverted pie pan.

Self-tapping screws. Self-tapping screws are useful in repair work, such as fixing squeaky wooden floor planks or stair treads. There are both flat-head and pan-head self-tapping screws. Because they are threaded the full length of the shank, self-tapping screws are suitable for many repair jobs where a nail won't hold. Self-tapping screws don't require pilot holes.

Drywall and all-purpose screws. These narrow, black screws look alike, but they are slightly different. All-purpose screws resist corrosion and have thicker shanks and coarser threads than drywall screws. Drywall screws are useful for any number of rough carpentry jobs, like hanging wallboard on wall framing. These self-tapping screws are installed with a power screw gun that drives the screw below the surface of the wallboard and disengages without damaging the wallboard. The tip of the screw gun holds the head of the screw with a magnet, which makes the job move quickly.

Any variable-speed electric drill with a hex-shank Phillips-head bit or adapter can be used to install them.

Particleboard screws. An ordinary wood screw has difficulty biting into the fibrous surface of composite materials. Particleboard screws, with coarsely pitched threads set at an angle, have the necessary holding power. A pilot hole isn't required.

Sheet-metal screws. These screws are different from wood screws because they are self-tapping and threaded their entire length. Sheet-metal screw heads may be flat, oval, or pan, and slotted or criss-crossed. They're used primarily for jobs like installing hinges, because their full-length threaded shaft provides maximum holding power.

Decking screws. Special screws have been developed for deck construction. Deck screws have greater holding power than nails and resist popping, a common problem with nails. Deck screws are made

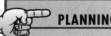

PLANNING POINTER

Choosing Deck Screws

The length of deck screw to use depends on the thickness of the deck boards. Remember that the screw should penetrate at least 1½ inches into the 2× deck joists. For example, if you use ¾-inch-thick deck boards, use a deck screw that is 2½ inches long; for ⁵⁄₄-inch-thick or 2 × 6 deck boards, use a screw that is 3 inches long.

of hardened steel and are specially treated to resist corrosion. They are available in sizes from small number 6 screws 3/8 to 2 1/2 inches long to large number 10 screws up to 3 1/2 inches long.

Most deck screws have a Phillips head and are meant to be installed with an electric screw gun with a matching tip. Since it takes a considerable amount of torque to drive the long screws, several different styles of screw-head slot types have emerged. You will find deck screws with square drives that require a matching square-drive tip for the screw gun. Before you purchase deck screws, check that you have the proper equipment to install the screws. Also purchase several extra tips for your screw gun since the tip will take a beating and eventually wear to the point that it will slip and damage the screw head.

Depending on the length of the screw and the hardness of the material you are driving it into, it may be helpful to predrill the screw holes using a drill bit that is slightly smaller in diameter than the screws. When you are building a deck with pressure-treated lumber, predrilling the screw holes is often necessary because the treated lumber is very dense and hard.

Nuts and Bolts

Nuts and bolts are a fastening system used to fasten wood and metal. The bolt is threaded or partially threaded with a blunt end and capped head, which is flat, round, and usually slotted, not crisscrossed. Square or hex nuts screw onto the threaded end to add strength to the bolt. If the surface is flexible, like thin wood or metal, a disk called a washer is used to protect it from the weight of the bolt head.

Nut and bolt lengths are described by two numbers. The first refers to the diameter; the second is the number of threads per inch. For example, a bolt listed as 3/4–14 is 3/4 inch in diameter and has 14 threads per inch. Lengths of nuts and bolts are listed in figure 2.3

NUT AND BOLT LENGTHS (INS.)

DIAMETER (INS.)	CAP SCREWS				BOLTS		
	ROUND HEAD	FLAT HEAD	HEXAGON HEAD	FILLISTER HEAD	MACHINE BOLT	CARRIAGE BOLT	LAG BOLT
1/4	1/2-2 1/4		1/2-3 1/2	3/4-3	1/2-8	3/4-8	1-6
5/16	1/2-2 3/4		1/2-3 1/2	3/4-3 3/4	1/2-8	3/4-8	1-10
3/8	5/8-3		1/2-4	3/4-3 1/2	3/4-12	3/4-12	1-12
7/16	3/4-3		3/4-4	3/4-3 3/4	3/4-12	1-12	1-12
1/2	3/4-4		3/4-4 1/2	3/4-4	3/4-24	1-20	1-12
9/16	1-4		1-4 1/2	1-4	1-30	1-20	
5/8	1-4		1-5	1 1/4-4 1/2	1-30	1-20	1 1/2-16
3/4	1-4		1 1/4-5	1 1/2-4 1/2	1-30	1-20	1 1/2-16
7/8			2-6	1 3/4-5	1 1/2-30		2-16
1			2-6	2-5	1 1/2-30		2-16
	Length intervals = 1/8 in. increments up to 1 in., 1/4 in. increments from 1 1/4 in. to 4 in., 1/2 in. increments from 4 1/2 in. to 6 in.				Length intervals = 1/4 in. increments up to 6 in., 1/2 in. increments from 6 1/2 in. to 12 in., 1 in. increments over 12 in.		Length intervals = 1/2 in. increments up to 8 in., 1 in. increments over 8 in.

Figure 2-3
Nuts and Bolts

Machine screws and bolts.   A machine screw is threaded its entire length. It has a screwlike head that's either slotted or crisscrossed, so a screwdriver is used, not a wrench. When it comes with nuts, a machine screw is called a stove bolt, and it's useful when assembling lightweight things.

A machine bolt is partially threaded with a flat end with a hex cap and nut, which stabilizes it and increases its holding power.

Carriage bolts.   Carriage bolts connect metal and wood members and are frequently used in the construction of furniture, metal components, and other assemblies. They have a round, smooth, nonslotted head forged at one end and a partially threaded shaft and a square nut.

U-bolts and J-bolts.   These are threaded steel rods in the shape of a U or J that are used as clamping devices and for hanging things.

Eye bolts.   These are specialty hooks in various shapes and configurations and are galvanized.

Lag bolts.   These heavy-duty bolts are threaded and fastened with a wrench because of their square or hex head. They're often used for outdoor furniture or assembling wooden components because of their strength. With softwood components, they're used with a washer to cushion the head when tightened.

Washers

Washers are doughnut-shaped metal disks used to spread the holding power of a fastener beyond the area covered by the nut or head on a bolt or screw. It is important to use washers on wood construction such as decks when you are joining two pieces of lumber. Washers are used on nut-and-bolt assemblies, or under the head of lag screws.

RULE OF THUMB

Avoiding Splits in Softwood

To avoid splits when using lag screws in softwood, drill a pilot hole that is one-half the diameter of the screw thread.

Washers are available in chromed steel, blued steel, or galvanized steel for corrosion resistance on exterior projects, or in a variety of decorative finishes for use on cabinets or furniture. Washers are available in many sizes to match the diameter of the bolt being used.

Washers are also used on metal projects. On tools or machinery that are subject to vibration or movement, a split-lock washer is used. A split-lock washer has a split or slot so that the washer is compressed as the bolt or screw is tightened. This pressure against the fastener head or nut prevents the fastener or nut from becoming loose due to repeated vibrations.

An external tooth washer serves a similar purpose to the split-lock washer. This washer has a serrated or scalloped outer edge. When the nut is tightened, the teeth grip the nut with tension to prevent the nut from loosening on the bolt.

Anchors

Anchors are used to fasten machinery, wrought iron, or other materials to concrete floors or slabs, and to fasten or hang materials or objects on walls. A wide choice of anchor devices is available to match your needs. Anchors are installed by drilling a hole to match the anchor diameter in the floor or wall. An anchor made of steel, plastic, fiber, or lead is inserted into the hole, and a screw or bolt is driven into the anchor. The anchors expand in the floor or wall material as the screw or bolt fastener is driven home.

To select the right anchor, describe your project to your hardware dealer. He will help you select the proper anchor for your particular project. Following are brief descriptions of various anchors for use on walls or floors and tips on installation.

Wedge anchors. Wedge anchors are used for anchoring materials to concrete floors. When the anchor is driven into the drilled hole, opposing wedges at both ends of the anchor are drawn tightly into the anchor and provide full-length expansion of the anchor against the sides of the hole. Depending on the object to be anchored, insert a screw or bolt into the anchor and tighten.

Plastic/nylon anchors. Like other anchors, plastic/nylon anchors are inserted into predrilled holes. The anchors are slotted in such a way that they expand and grip against the side of the hole when the screw is driven home through the center of the anchor. Choose a plastic/nylon anchor according to the thickness of the material to which you are anchoring (½-inch-thick wallboard, for example) and the weight of the object that will be supported by the anchor. Plastic/nylon anchors can be used to support towel racks, framed pictures, mirrors, or thermostats, or to anchor a glass shower door to a tiled wall.

Screw anchors. Screw anchors are inserted into holes bored into concrete, wood, plaster, or wallboard. The anchors can be made of plastic/nylon, metal, or even braided jute with a lead liner. After the anchors are installed, you can remove and replace the screws repeatedly while the anchor remains in place.

Zinc or aluminum drive anchors. These metal anchors are made with the same configuration as the more familiar molly bolt anchor: a bolt is inserted through a fluted collapsible shaft, and as the screw is turned, the shaft folds against the backside of the plaster, wallboard, or wood. In contrast to the molly bolt, the drive anchor has a pointed tip that can be driven through wallboard without predrilling a hole.

Lead anchors. Lead anchors have ribbed sides to grip the concrete or masonry firmly. Lead anchors are expansion-type anchors. Some are made in two halves; others are slotted on the bottom half. Both types expand to grip the hole edges as the screw or bolt is driven tight.

Machine bolt/screw anchors. Machine bolt/screw anchors are expansion-type anchors. Most have a small, tapered, steel wedge drawn into the body of the anchor by the machine screw. As the wedge moves into the anchor, it forces the body of the anchor to expand, locking it into the wall.

To install a machine bolt anchor, drill a hole of the same diameter as the anchor in the concrete. Insert the threaded end of the anchor into the hole so the top end of the anchor is flush with the surface. Some anchors, such as those made by Rawl, are secured in place with a setting tool provided with the anchors. Then insert the machine bolt/screw into the anchor and tighten.

Steel expansion anchors. Steel expansion anchors are made in one unit, similar to the molly bolt. The bolt extends through a steel shaft. The hole is drilled into the concrete, the expansion anchor is tapped home, and the steel shaft expands to grip the sides of the hole as the head of the bolt is tightened.

Zinc lag shields. Zinc lag shields are another type of expansion anchor. The body of the anchor is a single piece of soft metal

Figure 2-4 Zinc Lag Shields

SHIELDS FOR LAG BOLTS AND WOOD SCREWS (IN.)

LAG SCREW DIA. (IN.)	WOOD SCREW SIZES	DECIMAL EQUIV. (IN.)	LAG BOLT EXPANSION SHIELD			LEAD SHIELD FOR LAG BOLT OR WOOD SCREW	
			A	L SHORT	L LONG	A	L
	6	.138				1/4	3/4–1 1/2
	8	.164				1/4	3/4–1 1/2
	10	.190				5/16	1–1 1/2
	12	.216				5/16	1–1 1/2
1/4	14	.250	1/2	1	1 1/2	5/16	1–1 1/2
	16	.268				3/8	1 1/2
	18	.294				3/8	1 1/2
5/16	20	.320	1/2	1 1/4	1 3/4	7/16	1 3/4
3/8	24	.372	5/8	1 3/4	2 1/2	7/16	1 3/4
1/2		.500	3/4	2	3		
5/8		.625	7/8	2	3 1/2		
3/4		.750	1	2	3 1/2		

with several slots cast into the shaft of the shield. The shield is forced to expand as the lag bolt is screwed down into the center of the anchor. The soft metal expands and grips the surrounding wall. Sizes of common shields are given in figure 2.4. As with other anchors, drill a hole in the concrete and insert the lag shield. Tap the shield with a hammer to drive it flush to the surface. Then insert the lag bolt and tighten it.

Hollow wall anchors. Hollow wall anchors, also known as molly bolts, can be used to anchor objects to wallboard, plaster, paneling, or hollow-core doors. The units consist of a collapsible threaded shaft with a screw or bolt driven through the center of the shaft. Check the package label to be sure the wall anchor is the right model and size for your project. To install this type anchor, drill a hole using the drill size suggested on the directions. Then tap the wall anchor into the hole, and use a screwdriver to tighten the screw until you feel resistance, which means the metal shaft has folded against the backside of the wallboard or plaster. Sizes of hollow wall anchors are given in figure 2.5.

Toggle bolts. A toggle bolt consists of a bolt threaded onto spring-loaded folding "wings" or toggles. Toggle bolts can be used to secure objects to wallboard, plaster, block, or hollow tile. To install a toggle bolt, drill a hole to match the diameter of the toggle bolt. Fold the toggles toward you and push the bolt through the hole in the wall. Once inside the hollow cavity of the block or wall, the toggles will spring open, and as the bolt is tightened, the toggles will lock against the backside of the material. Sizes of toggle bolts are given in figure 2.6.

Figure 2-5 Hollow Wall Anchors

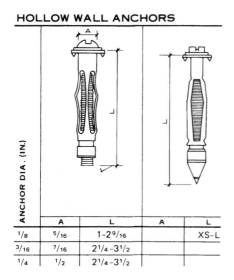

HOLLOW WALL ANCHORS

ANCHOR DIA. (IN.)	A	L	A	L
1/8	5/16	1–2 9/16		XS-L
3/16	7/16	2 1/4–3 1/2		
1/4	1/2	2 1/4–3 1/2		

Figure 2-6 Toggle Bolts

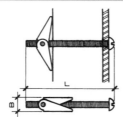

SPRING WING TOGGLE BOLTS (IN INCHES)							
DIAMETER	$1/8$	$5/32$	$3/16$	$1/4$	$5/16$	$3/8$	$1/2$
DECIMAL EQUIV.	.138	.164	.190	.250	.313	.375	.500
B	.375	.500	.500	.688	.875	1.000	1.250
L	2 – 4	$2 1/2$ – 4	2 – 6	$2 1/2$ – 6	3 – 6	3 – 6	4 – 6

Special large toggle bolts with 11-inch spans will spread the load over a greater area of the ceiling, and will support much heavier loads than normal-size toggles. They can be used to hang heavy plants or fixtures from the ceiling.

Tie-wire toggle bolts are used to secure tie wires for suspended ceilings beneath existing tile, metal, wallboard, or plaster ceilings. They have a washer and hex nut on the shaft to tighten the toggle bolt against the old ceiling material. They have a spade tip or end with a $7/32$-inch hole through which tie wires are run to support the metal grids of a suspended acoustical ceiling.

Metal Framing Connectors

During the Depression, metal wood-connecting devices were developed to help expand the construction industry and to create new jobs. Some of these metal connectors are visible in the trusses in your roof, in joist hangers to connect headers and joists at stairways, or under outdoor decks. Metal framing connectors not only result in a stronger wood joint, especially when used as an alternative to nailing where wood members are connected at right angles, but also solve the problem of connecting wood framing to other materials, such as connecting deck posts to concrete base piers.

Use metal rafter tie-downs or hurricane anchors to secure ceiling joists or roof trusses to the top plate of walls. This connector reinforcement makes the roof more resistant to damage from high winds, storms, or even earthquakes. Use U-post bases and post anchors to fasten structural members to concrete foundation piers. Use metal stair angle support straps to fasten stair stringers to joists; joist supports to fasten floor joists at right angles to rim joists; post beam caps to fasten the top of deck posts to the joists; staircase angles to fasten steps to stair stringers; and all-purpose anchors that can be bent to serve almost any wood-connecting project. Some of the more common metal framing connectors are shown in figure 2.7.

There are dozens of other applications for metal framing connectors. You can pick up a manufacturer's brochure listing the many types available at most hardware stores or home centers.

Staples

A staple is a nail-like fastener installed with either a hand or electric staple gun that forcefully shoots the staple into the surface of the material. Staples are suitable for fastening or tacking thin materials, such as carpeting, window screening, ceiling tiles, weather stripping, plastic drop-cloth sheeting, insulation, and other materials. There are primarily six sizes of staples: $1/4$ inch, $5/16$ inch, $3/8$ inch, $1/2$ inch, $9/16$ inch, and $17/32$ inch.

Figure 2-7 Metal Framing Connectors

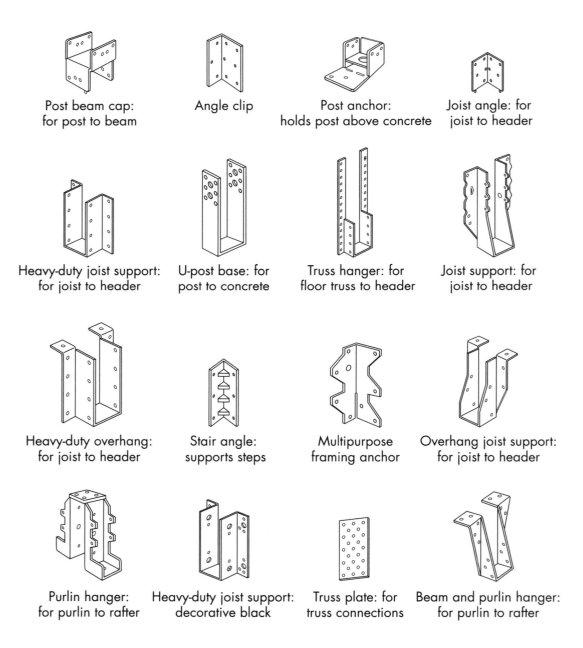

Post beam cap:
for post to beam

Angle clip

Post anchor:
holds post above concrete

Joist angle: for
joist to header

Heavy-duty joist support:
for joist to header

U-post base: for
post to concrete

Truss hanger: for
floor truss to header

Joist support: for
joist to header

Heavy-duty overhang:
for joist to header

Stair angle:
supports steps

Multipurpose
framing anchor

Overhang joist support:
for joist to header

Purlin hanger:
for purlin to rafter

Heavy-duty joist support:
decorative black

Truss plate: for
truss connections

Beam and purlin hanger:
for purlin to rafter

Rivets

A rivet is a nail-like fastener that bonds together thin materials, such as sheet metal, gutters, canvas, and leather. Rivets are installed by a rivet tool that flattens the flanges of the rivet on the surface, compressing the rivet together. Rivets are the fastener of choice when repairing metal lawn furniture, aluminum doors and windows, bicycles, television antennas, and for any other jobs where thin materials are fastened.

Rivets come in steel and aluminum in sizes ranging from $\frac{1}{8}$ to $\frac{3}{16}$ inch. Use steel rivets for heavy-duty jobs and aluminum rivets for lightweight materials. The wider its diameter, the more fastening power a rivet has. To decide the grip range, or how long the rivet should be, use the thickness of the materials being fastened. The thinner the material, the shorter the rivet and, thus, the smaller the grip. When fastening soft materials like fabric, add a washer to spread the weight of the fastener.

RULE OF THUMB

Installing Rivets

Install the rivet at least one and a half times its diameter from the edge of the piece it is in. Figure that a properly riveted joint will have three-fourths the strength of the pieces it joins together.

Adhesives and Glues

Adhesives and glues were once commonly used only for interior applications, but modern technology has been used to develop weatherproof adhesives, such as the new line of deck adhesives. Following is a review of many types of adhesives and glues, and their applications.

Adhesives

Most of us are familiar with tile, carpet, and floor-covering adhesives. But there are many other adhesive applications that make work easier for the handyman. For example, wallboard adhesives can help bridge over minor irregularities in framing, and can be used instead of screws or nails to secure wallboard to wood studs or framing. Use wallboard adhesives to fasten wallboard to concrete, to fasten drywall to the lightweight framing used for pocket doors, or to secure a new wallboard skin over deteriorated plaster walls or ceilings.

Construction adhesives applied between floor joists and plywood subfloors can help to eliminate floor squeaks, reduce the number of nails needed, and build a stiffer floor. Use adhesives to install countertops on kitchen cabinets or bathroom vanities.

When finishing rooms in a basement, use adhesives to secure wall sole plates to concrete floors, or to secure corner studs or backing to concrete walls. Use adhesives to secure nailing blocks or framing to cover and conceal steel I-beams. Exterior deck or docking adhesives are strong enough to resist direct water contact and eliminate popped nails that might injure bare feet. In short, there is a construction adhesive to help you perform almost any home project. Check out the assortment at your home center.

Glues

Although many all-purpose glues are available, there is always a best glue choice, depending on the project. It is best to buy a glue that is formulated for a specific purpose rather than try the "one size fits all" approach. Following is a brief review of glues and their uses.

Aliphatic glue.   Aliphatic (yellow) glue is a good choice for woodworking projects. Aliphatic glues are available in plastic squeeze bottles. Clean up spills with a damp cloth. Aliphatic glues are quick-setting, but glue projects should be clamped for maximum adhesion. Clamping time is about 2 hours.

Contact cement.   Contact cement is usually used for bonding plastic laminates to countertops or walls, but it can also be used to join plywood, leather, metal, or rubber, and to attach wood veneers or edge banding on furniture or cabinet projects. Use a brush or roller to coat both surfaces to be joined, and let the cement set until it is dry to the touch before joining the two. Contact cement is available in both solvents and water bases. Solvent cements are flammable, so observe the label warnings.

The term *contact* means that the two work pieces will bond tightly at the point of contact, so you must be careful to avoid touching the two pieces together until they are perfectly positioned. Wood lath, dowels, or brown paper is often placed on one surface after the cement is dried, then pulled away carefully as the two pieces are mated. Contact cements do not require clamping, but the surface should be pressed with a roller to ensure full contact between the two pieces.

Cyanoacrylate glue.   Cyanoacrylate, or instant, glue has many household uses, such as repairing broken pottery or ceramics. It can be used on nonporous materials only; the glue cannot be used for gluing wood or paper projects. The original cyanoacrylate glue was called Crazy Glue. The solvent for cyanoacrylate glue is acetone or fingernail polish remover.

Epoxy cement.   Epoxy cement is a two-part adhesive that must be mixed carefully to achieve full bonding strength. Epoxy cements are not the best choice for woodworking projects, but can be used to bond tile, metal, or other materials to wood. Most epoxy projects do not require clamping. Mix the resin and hardener together in small quantities because the cement will set in the container. Check setting time on the product label.

Plastic resin glue.   Plastic resin glues are sold in dry powder form and must be mixed with water. Because plastic resin glues set in 2 hours or less, you should mix only small amounts that can be used quickly. These glues can be used for wood projects, but the glue line is brittle. For best results, the parts or edges should be planed to fit tightly together. Leave clamps on for 12 to 14 hours.

Resorcinol glue.   Resorcinol glues are extremely waterproof and can be used for outdoor projects or even for boats. Resorcinol glues are two-part products—resin and a catalyst—that set quickly and should be mixed only in small amounts. Resorcinol glues will fill small gaps in the joint. Clamping time can be 14 hours or longer.

Silicone glue.   Some silicone products are offered both for adhesive use and for caulking. For example, you can use these products both to reglue a loose ceramic tile and to caulk the joints. They are available in plastic squeeze tubes. Check the product label to see if the silicone product is recommended for your particular project.

White carpenter's glue. White carpenter's glue, also known as polyvinyl acetate, is used for a wide variety of carpentry and hobby projects. The glue is sold in plastic squeeze bottles, is easy to use, can be cleaned up with water, and sets quickly. Clamp parts immediately when the glue is spread. Clamp time can be up to 2 hours.

Clear airplane glue. Clear airplane glue is sold in a squeeze tube and can be used for many hobby projects, including model building. For other hobby glues, check in hobby or fabric shops and review their entire stock to find the right glue for your needs.

3 Rough Carpentry and Finish Carpentry

The term *carpentry* once referred to the skills needed for building things with wood. In today's terminology, carpentry also refers to the assembly of many different building components, including those made of wood, plastic, metal, and other materials.

The skills of carpentry are generally divided into two levels, rough carpentry and finish carpentry.

Rough carpentry refers to those skills necessary to set forms for concrete and to lay out and frame the floors, exterior shell, and interior walls of a building. Rough or framing carpenters frame the window and door openings in exterior walls, set roof trusses, and install subfloor and wall and roof sheathing plywood.

Finish carpentry refers to the skills necessary to do such finishing work as installing interior trim, hardwood flooring, closet shelving, doors, cabinets, and countertops.

In this chapter we will discuss those carpentry skills most often needed by the homeowner, both for rough and finish carpentry.

Carpentry Tools

For carpentry work you will need tools that measure, cut, drill, smooth, and connect wood and other building materials, such as those described below. You will also need scaffolds and ladders.

Tools for measuring wood. Measuring tools include a quality 12-foot measuring tape, a carpenter's square, and a machinist's square.

Tools for basic cutting. Cutting tools include a handsaw, a circular or power saw, and a saber saw for scroll or curved cutting. As carpentry skills increase, the homeowner may wish to add to this list stationary power tools, such as a table saw, a radial arm saw, or a scroll saw.

Tools for basic drilling. The electric drill has replaced the once common hand brace and bits, although many carpenters still prefer a hand brace for some jobs when working with furniture or cabinets. Today the standard drilling tool is a reversible $\frac{3}{8}$-inch electric drill and a variety of bits. Also available are hole saws, useful for such tasks as boring a hole in a door for installing a lockset.

Safety Tips for Using Ladders and Scaffolds

- For ladder safety, set the base of the ladder about one-fourth the ladder's length from the house. For example, set a 16-foot ladder's base about 4 feet away from the wall. Avoid pulling extension cords or water hoses about while on a ladder. Any item that can tangle your feet and cause a trip hazard should be avoided. Use cordless power tools when working on ladders.

- When buying stepladders, check their duty rating or weight limits listed on the ladder's label. A Type I ladder will support 300 pounds; a Type II ladder, 250 pounds; and a Type III ladder, 200 pounds. Note that the weight limits include both the weight of the person and any tools or materials carried on the ladder.

- When using a stepladder, spread the legs completely and lock the folding braces out so they prevent the legs from folding. Check the stepladder for cracks, loose hardware, or broken steps. When using a metal ladder, stay well away from any overhead electric power lines.

- Do not use chairs or other objects as scaffolds or ladders. You can make a sturdy work platform by using two sawhorses with planks between. Check scaffold planks carefully. Professional riggers know that a plank that will flex or bend slightly under your weight is safer than a stiff plank that won't bend but may break without warning. Test any plank at low level—that is, set one end of the plank on the ground, the other end on the lower step of a ladder, then walk on the plank—before trusting it on a high assignment where a fall could prove dangerous.

- For high work where a work platform is required, rent knockdown steel scaffolding. These tubular steel scaffolds can be assembled in any configuration and to reach any reasonable height. The braces and scaffolds can be joined quickly via spring clips or steel pins, so assembly and disassembly are quick and easy.

- When working on either ladders or scaffolds, wear clothes that will not be a climbing hazard. Avoid loose-fitting clothing that may become hooked on nails or ladders. Don't wear rings or other jewelry when climbing; rings can catch on a nail and injure a finger. To provide ankle support, wear over-the-ankle boots and check to be sure that shoestrings are securely tied. Don't try to save time by overreaching the ladder or scaffold; rather than lean far out, climb down and move the scaffold.

Tools for smoothing wood. Smoothing tools include hand planes, sanders, and wood rasps. For furniture or cabinet work, one needs an electric jointer/planer and a variety of special-use power sanders.

Tools for connecting wood or other materials. Connecting tools include a 16-ounce carpenter's hammer and nails. A selection of screwdrivers, including blade or slot and Phillips tips, in a variety of sizes is needed for fastening with screws. A variety of power screwdrivers and screwdriver bits that can be used with a ⅜-inch drill round out the list of connecting tools.

Scaffolds and Ladders

Many home repair jobs can be done while standing on a ladder. The chief drawbacks to ladders are that you must move them frequently, and long hours standing on ladder rungs can tire your feet. For installing siding or painting, you can rent ladders, ladder jacks, and planks. Ladder jacks have brackets that fit over a pair of rungs, forming a flat support at right angles to the ladders. Planks can then be laid between the ladder jacks on a pair of ladders, providing a work platform that will permit you to work on long sections of walls.

Rough Carpentry

General Layout

Today's houses are built using modular layout to accommodate sheathing, insulation, and wallboard panels. The framing members—studs for the walls, joists for the floors and ceilings—are all laid out or spaced to fit 4 × 8 panels. Thus, the framing members—either studs, joists, or roof trusses—

Table 3.1 Number of Studs or Joists per Running Foot

16" O.C. Spacing

Running Feet	4	4.5	5	5.5	6	6.5	7	7.5	8	8.5	9	9.5	10	10.5	11	11.5	12
Number of joists or studs	4	5	5	6	6	6	7	7	7	8	8	9	9	9	10	10	10
Running Feet	13	13.5	14	14.5	15	15.5	16	16.5	17	17.5	18	18.5	19	19.5	20	20.5	21
Number of joists or studs	11	12	12	12	13	13	13	14	14	15	15	15	16	16	16	17	17

24" O.C. Spacing

Running Feet	4	4.5	5	5.5	6	6.5	7	7.5	8	8.5	9	9.5	10	10.5	11	11.5	12
Number of joists or studs	3	4	4	4	4	5	5	5	5	6	6	6	6	7	7	7	7
Running Feet	13	13.5	14	14.5	15	15.5	16	16.5	17	17.5	18	18.5	19	19.5	20	20.5	21
Number of joists or studs	8	8	8	9	9	9	9	10	10	10	10	11	11	11	11	12	12

RULE OF THUMB

Justifying Framing 24 inches o.c.

Framing 24 inches o.c. uses about one-third less framing lumber than framing 16 inches o.c. This conserves valuable forest resources, plus reduces housing costs to the consumer by reducing both material and labor costs.

An exception to the rule. Where roof trusses are set 24 inches o.c., the weight of ceiling insulation can cause ½-inch-thick wallboard panels to sag between the trusses. Use the more rigid ⅝-inch-thick wallboard panels for ceilings where trusses or joists are set 24 inches o.c.

(for 16-inch o.c. spacing) or 22½ inches wide (for 24-inch o.c. spacing).

Plywood or insulating rigid foam panels are used as sheathing for floors, walls, and roofs. These materials are all made in 4×8 panels. Wallboard panels are also manufactured in 4-foot increments, always 4 feet wide and either 8 or 12 feet long.

For decades 16-inch o.c. spacing was the standard in residential construction for both stud and joist framing. However, industry tests have proven that 24-inch o.c. spacing provides adequate structural strength, so 24-inch o.c. spacing is common for today's housing, whether for wall studs, ceiling joists, or roof trusses. (See table 3.1.)

Studs—Wall Framing

Traditionally, the word *studs* refers to the rough lumber, usually 2×4s, used to frame the walls of a house. To permit installation of thicker wall insulation, the building industry has moved toward use of 2×6s for exterior wall framing, while still using 2×4s for framing interior wall partitions. Building code requirements for wood studs in bearing and nonbearing walls are shown in table 3.2.

are laid out so they are spaced at either 16 inches or 24 inches on center (o.c.), measured from the center of one stud or joist to the center of the next stud or joist. Also, fiberglass insulation blankets or batts are made to fit between the framing members, so the insulation is either 14½ inches wide

Table 3.2 Building Code Requirements (Stud-Grade Lumber)

	Stud Size	2×4	2×6
WOOD STUDS IN BEARING WALLS			
Wall height not to exceed:		10'	10'
Spacing (o.c.) with only ceiling and roof load		24"	24"
Spacing (o.c.) 1 floor, ceiling and roof load		16"	24"
Spacing (o.c.) 2 floors, ceiling and roof load		NA[a]	16"
WOOD STUDS IN NONBEARING WALLS			
Wall height not to exceed:		14'	20'
Spacing (o.c.)		24"	24"

[a]NA—not available

Also, as lumber prices have gone up, the industry has moved toward wider use of steel studs rather than wood studs for residential framing. Steel studs are available in either 2 × 4 or 2 × 6 size. To work with steel studs, you need metal shears for cutting the steel, drills for penetrating the steel, and pop-rivet or die-cut tools for securing the steel framing together. A drywall screw gun and self-tapping drywall screws are also needed for installing the drywall or wallboard finish. For the do-it-yourselfer who has limited access to steel-working tools, wood studs will probably prove easier to work with than steel.

Joists—Floor Framing

When building a floor one must keep in mind the live loading, that is, the joist strength necessary to support both the weight of the floor itself (called the dead load) and the

RULE OF THUMB

Advantages of Steel Studs

Steel studs are absolutely straight, so they are excellent for use on kitchen soffits or anywhere a very straight wall or soffit is desirable. Unlike wood studs, steel studs do not warp or shrink, so you will have fewer nail or screw pops when using steel studs.

weight of furniture and people that will occupy the floor space (called the live load). The required live loading for residential floors is 40 pounds per square foot (psf). The joist under the live load must not deflect more than $1/360$ of the span. For example, a joist that is 12 feet long can sag or deflect no more than $4/10$ inch at the center under the live load. Table 3.3 gives allowable

Table 3.3 Floor Joist Span   Maximum spans given in feet and inches.

FLOOR JOISTS—40 PSF LIVE LOAD, 20 PSF DEAD LOAD, 1/360
All rooms except sleeping rooms and attic floors (maximum 1.5" lightweight concrete)

Size	Spacing	Grade									
Inches	Inches on Center	Dense Select Structural	Select Structural	Nondense Select Structural	No. 1 Dense	No. 1	No. 1 Nondense	No. 2 Dense	No. 2	No. 2 Nondense	No. 3
2 × 6	12	11-4	11-2	10-11	11-2	10-11	10-9	10-11	10-9	10-3	8-6
	16	10-4	10-2	9-11	10-2	9-11	9-9	9-11	9-6	9-1	7-5
	24	9-0	8-10	8-8	8-10	8-8	8-6	8-5	7-9	7-5	6-0
2 × 8	12	15-0	14-8	14-5	14-8	14-5	14-2	14-5	14-2	13-6	10-10
	16	13-7	13-4	13-1	13-4	13-1	12-10	13-1	12-4	11-9	9-5
	24	11-11	11-8	11-5	11-8	11-3	10-8	10-10	10-0	9-7	7-8
2 × 10	12	19-1	18-9	18-5	18-0	18-5	18-0	18-1	17-0	16-1	12-10
	16	17-4	17-0	16-9	17-0	16-4	15-8	15-8	14-8	13-11	11-1
	24	15-2	14-11	14-7	14-1	13-4	12-10	12-10	12-0	11-5	9-1
2 × 12	12	23-3	22-10	22-5	22-10	22-5	21-7	21-7	19-10	19-1	15-3
	16	21-1	20-9	20-4	20-3	19-6	18-8	18-8	17-2	16-7	13-2
	24	18-5	18-1	17-9	16-6	15-11	15-3	15-3	14-0	13-6	10-9

Table 3.4 Floor Joist Notching and Drilling

	Maximum Depth or Diameter			
	2 × 6	2 × 8	2 × 10	2 × 12
Notch at joist end[a]	1⅜"	1⅞"	2⅜"	2⅞"
Notch in outer third of joist[a]	⅞"	1¼"	1½"	1⅞"
Hole anywhere in joist[b]	1½"	2⅜"	3"	3¾"

[a]No notching allowed in center third of joist.
[b]All holes must be at least 2 inches from joist edge.

spans for each species of lumber, assuming a live loading of 40 pounds per square foot with a deflection of not more than ⅟₃₆₀.

Because of the effects of moisture and weather extremes on wood, you should use solid dimension lumber for joists in outdoor structures such as decks. Floor joists are set at either 16 inches o.c. or 24 inches o.c. Lumber for floor joists is usually selected from among the various dimension lumber sizes, such as 2 × 6, 2 × 8, 2 × 10, or 2 × 12. The lumber you choose for floor joists depends on two things: the species of lumber used and the width or span of floor the lumber must support.

For interior joist use, the industry has developed manufactured floor joists that are straight, strong, and lightweight for easy installation. These manufactured joists have parallel edges or rails of grooved 2 × 2s, with a solid plywood web between the edge rails. Manufactured joists provide a level and squeak-free floor without the shrinking, warping, or twisting that may occur with solid wood joists.

For the floor joist to maintain its designed strength, care must be taken when notching the joist or drilling holes in it. Building codes have strict requirements covering such practices. Generally, no notches or large holes can be made in any joist in the center third of the span. Table 3.4 gives the maximum depth or diameter that a joist can be notched or drilled and still maintain its full load-bearing ability.

Shingle and Shake Layout

The rule of thumb for spacing roof shingles is, the steeper the roof slope, the wider the spacing can be. For example, on a roof with a 5/12 slope (the roof rises 5 inches for every 12 inches of horizontal travel), the exposure or spacing for a 24-inch wood shingle can be 7½ inches, but for a more gentle 3/12 slope, the shingle exposure can be only 5¾ inches. The obvious reason for this difference in allowable exposure is that water will run more quickly off the steeper 5/12 roof slope than it will off the 3/12 roof slope, so the shingle exposure spacing can be greater for the steeper 5/12 roof slope (see table 3.5; the spacings listed are the maximum allowable).

For all shingles or shakes, you must measure the distance in inches from the roof ridge to the roof edge, then divide this distance by the maximum allowable exposure to find the number of rows, or courses, of shingles you will have. For uniform spacing—that is, to avoid having one course or row of shingles that does not have

Table 3.5 Maximum Exposure Shingles and Shakes

Roof Slope	Shingle and Shake Length		
	16"	18"	24"
5/12 & up	5"	5½"	7½"
4/12	4½"	5"	6¾"
3/12	3¾"	4¼"	5¾"
On vertical wall	6–7½"	7–8½"	8–11½"

the same exposure as the other courses—the number of shingle courses must come out even. Adjust the exposure of the shingles so that you have an even number of courses, all showing the same inches of exposure.

For example, if the distance (width) from the ridge to the edge is 150 inches, and we divide 150 inches by the 7½-inch maximum allowable exposure, we would have 20 courses of shingles, all showing the same exposure for good appearance. Suppose that the roof width from ridge to edge were 140 inches. We would divide 140 inches by 7½ inches (maximum allowable exposure) and get 18.66, or 18⅔ courses. To eliminate the two-thirds course, we would divide the 140 by 19 courses and get 7.37, or 7⅜ inches. We would then install 19 rows of shingles at 7⅜-inch exposure.

Paneling

In the past, interior plywood wall paneling had a face ply, or outer layer, of real wood. Because the ply was peeled from a log, the grain and knots in the wood varied from panel to panel. This variation in appearance is still true today in high-quality authentic wood paneling. To achieve a pleasant panel-to-panel uniformity, stand real wood panels against the wall and check how the grain and other characteristics look when viewed next to each other.

For example, if you have a few panels that have a really wild or prominent grain pattern, you may want to group them together so the wild grain panels are all on one wall, rather than setting one wild-grain pattern between two panels with a more subdued grain pattern. The point is to lay out or arrange the panels so there is continuity and flow as you look at the walls, rather than have contrasting panels adjoining each other.

Modern interior paneling has a base of hardboard or low-cost wood plies, and a vinyl surface coating with a printed grain pattern that is alike on all panels, so there is no need to arrange the panels to avoid pattern and grain clash.

When laying out interior wall paneling, be sure to plan your layout so all panel edges fall on wall studs and you can thus securely nail the panel edges.

Most wall paneling has grooves running parallel to the long edges of 4×8 panels. Where possible, lay out the 4×8 panels so that the grooves fall over the studs in the walls, then drive the paneling nails in the grooves so they are concealed. If there is no groove at the stud location, drive paneling nails in a spot where there is a knot or wood grain. The nail heads will be more easy to conceal when they are driven into irregular areas, such as knots or wild-grain marks.

Siding Layout

When laying out wood or hardboard lap siding, you should first measure and mark off a story pole. To make a set of story poles, which are used as a layout guide, cut two straight pieces of 1×4 long enough to span the area the lap siding will cover, from the foundation to the roof soffit, or

RULE OF THUMB

Using Panel Adhesives

Where possible, use panel adhesives instead of nails. Adhesives present superior holding power while eliminating blemishes created when nails are used.

overhang. Measure the distance from the bottom edge of the first strip of siding to the top edge of the top strip of siding, and mark these two boundaries on the two opposite ends of both story poles. Let us assume that this distance is 100 inches.

Now measure the width of the siding boards and allow something for the overlap at the siding joints. For example, assume that you are installing 1 × 12 rough-sawn cedar siding that has an actual width of 11½ inches, and you will overlap the siding approximately 1 inch at each joint. Divide the length of the area the siding will cover, or 100 inches, by 10½ inches (11½ minus 1), and you will see that you will need 9½ courses, or rows, of siding to cover the area. If we decide we do not want only a half-course at the top, we could instead plan on leaving 10 inches of siding per board (to the weather) and have 10 courses. Decide how much width you will leave to the weather, and mark the two story poles at these intervals. Then temporarily nail the two story poles at the two opposite corners of the wall. Stretch a chalk line from the bottom edge mark on one story pole to the bottom edge mark on the opposite story pole. Install the first course of siding and move the chalk line up to the second mark. These two marked story poles will be the guides you will follow to keep the courses of siding straight and level.

When installing siding, it is necessary to observe proper nailing methods. Figure 3.1 shows proper nailing methods for most of the popular siding designs. Failure to observe these nailing methods can result in split or buckled siding.

Prefinished siding panels such as aluminum and vinyl clip together, so there is no opportunity to adjust the spacing of the siding. To keep the siding courses level, it is useful to use a pair of story poles or a chalk line when installing prefinished aluminum or vinyl siding.

RULE OF THUMB
Cleaning Siding

Use commercial wood deck cleaners to remove oxidized paint from aluminum siding. Use a shop-grade hand cleaner to remove adhesives or caulks from vinyl siding.

Roof Carpentry and Layout

Aside from building cabinetry or fine furniture, the carpentry task that requires the most knowledge may be framing a roof. For example, you must cut precise angles to fit rafters to ridge boards, which run down the center of the roof at the peak. Jack rafters, used where two roof sections intersect, and hip rafters, used for a hip roof, vary in length and may have to be cut with hard-to-calculate compound angles. To learn how to make these angle cuts and lay out rafters, buy a carpenter's framing square and learn to use it. An instruction booklet included with the square will show you how to use the square to cut the needed angles and to step off the rafters for length.

Roof Framing

Figure 3.2 gives a graphic view of how complicated framing a roof may be. Most of us will never undertake building an entire roof, and the job should not be undertaken without considerable study of the subject. Keep in mind that the strength of the roof depends entirely upon the fact that each angle cut is precise, so that rafters meet other rafters or ridge boards in such a manner that the angled end fits tightly against the joined member, and the joined members meet or press evenly along their entire joining surfaces.

Figure 3-1 Proper Nailing Methods

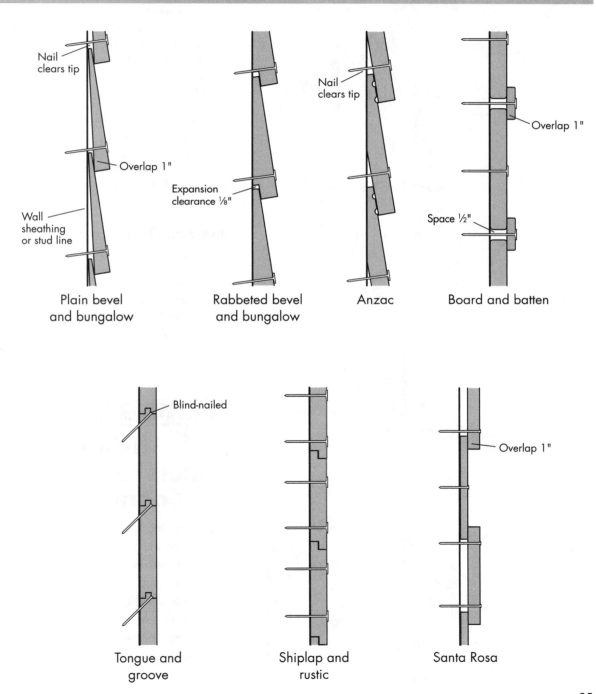

Plain bevel and bungalow

Rabbeted bevel and bungalow

Anzac

Board and batten

Tongue and groove

Shiplap and rustic

Santa Rosa

Figure 3-2
Roof Framings

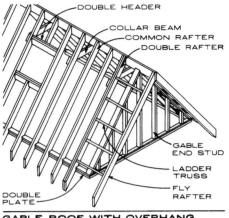

DOUBLE HEADER
COLLAR BEAM
COMMON RAFTER
DOUBLE RAFTER
GABLE END STUD
LADDER TRUSS
FLY RAFTER
DOUBLE PLATE

GABLE ROOF WITH OVERHANG

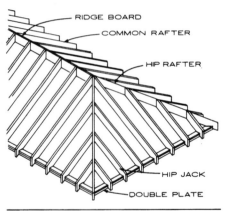

RIDGE BOARD
COMMON RAFTER
HIP RAFTER
HIP JACK
DOUBLE PLATE

HIP ROOF

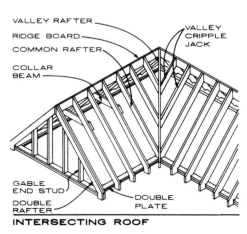

VALLEY RAFTER
RIDGE BOARD
COMMON RAFTER
COLLAR BEAM
VALLEY CRIPPLE JACK
GABLE END STUD
DOUBLE RAFTER
DOUBLE PLATE

INTERSECTING ROOF

Truss Framing

The good news for the consumer is that we no longer need to master the mystery of carpenter's math for framing roofs. Over the past several decades, the building industry has gone almost exclusively to roof truss construction. Companies that specialize in truss construction will custom-build trusses for almost any building, from small storage sheds and garages right up to the largest houses. These custom-built roof trusses are available in any length, and can be built with any roof pitch desired. Truss framing is shown in figure 3.3.

Window/Door Layout

Before framing door or window openings, you must decide how large the door or window will be, and how large the rough opening (sometimes abbreviated r.o.) should be. The rough opening should be large enough so that the door or window unit can be moved freely and plumbed and leveled

RULE OF THUMB

Using Metal Framing Connectors

For framing special projects, don't overlook metal framing connectors (described in chapter 2), which can be put together in various ways to build a storage shed and other structures. Look for metal framing connectors at any home center.

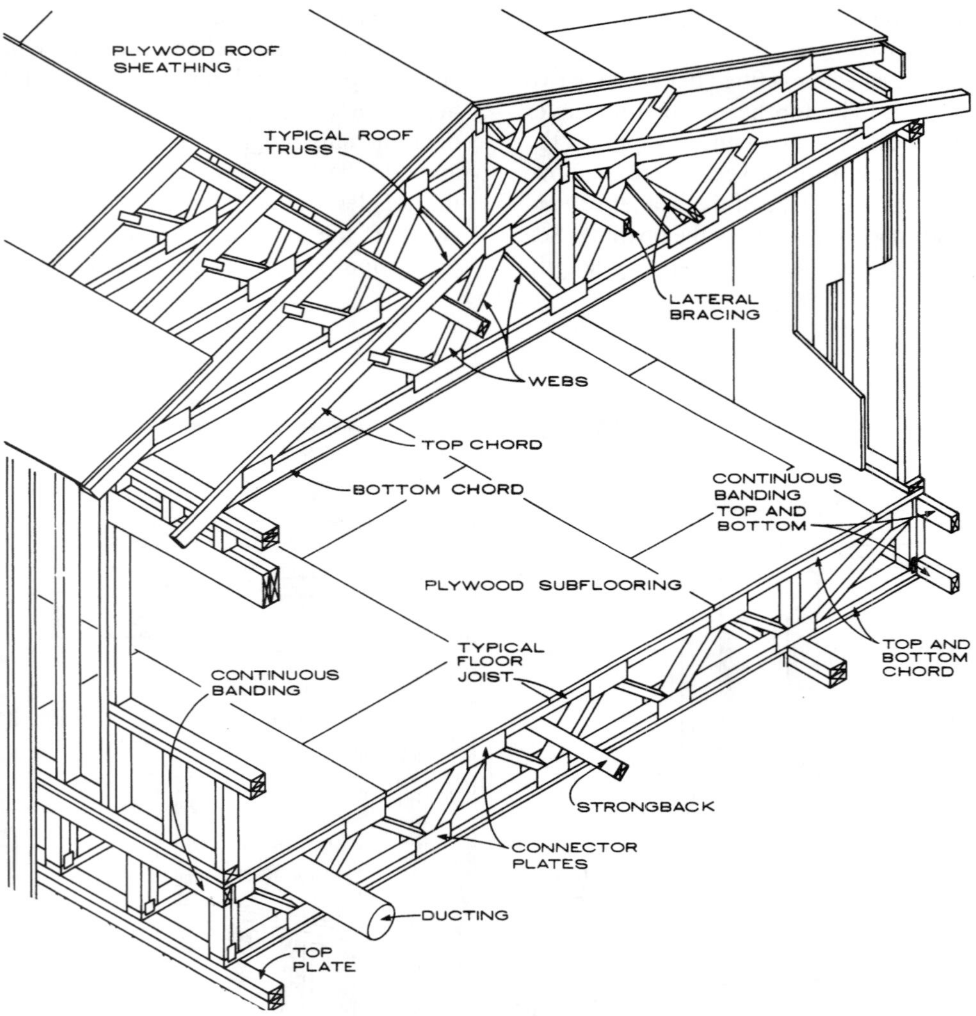

Figure 3-3
Truss Framing

PLYWOOD ROOF
SHEATHING

TYPICAL ROOF
TRUSS

LATERAL
BRACING

WEBS

TOP CHORD

BOTTOM CHORD

CONTINUOUS
BANDING
TOP AND
BOTTOM

PLYWOOD SUBFLOORING

TYPICAL
FLOOR
JOIST

TOP AND
BOTTOM
CHORD

CONTINUOUS
BANDING

STRONGBACK

CONNECTOR
PLATES

DUCTING

TOP
PLATE

within the rough opening. Tapered wood shims are used to ensure the door or window unit is plumb and level, and to nail the unit in the opening.

For windows, the rough opening should be about 1½ inches larger both in height and width than the outside measurements of the window unit (measured from the outside edge of the window frame).

For interior doors, the rough opening can be 2⅛ inches wider than the actual width of the door, and 2⅝ inches higher than the height of the actual door (not including the frame).

The framing members used in door and window layouts are shown in figure 3.4. For greater strength, the sides of the rough opening are two full-length studs, called king studs, set at the proper width. (Spacing depends on the width of the door or window unit.)

Next, a header is fitted between the two king studs. The top header or lintel over the door or window opening supports the weight of the roof or second story above the opening. The size of the header needed depends upon the width of the opening. See table 3.6 for proper header size by length.

Note that the size of lumber used for the header is doubled; that is, two members of the size noted are spiked together, with a filler strip of ½-inch plywood, to form the header. For example, for a 3–0 (3-foot) opening, spike together two 3-foot 2 × 4s, and add the ½-inch plywood filler so that the header is the same thickness as the studs. If the studs are 2 × 4s, the actual dimensions are 1½ inches × 3½ inches. So

Figure 3-4 Framing Doors and Windows

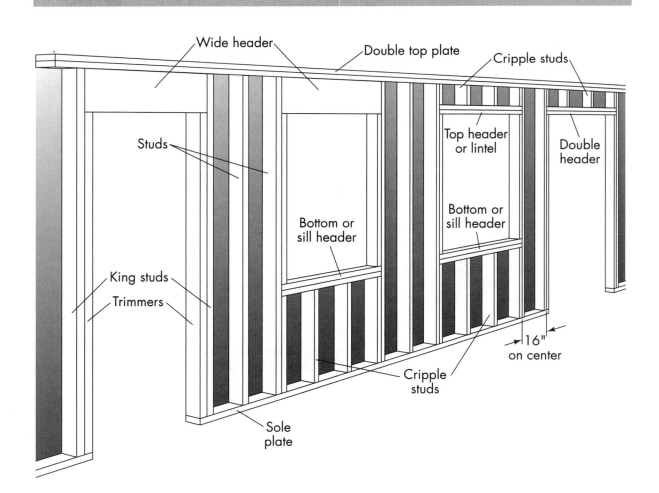

Table 3.6 Wood Header Spans

| | (Douglas Fir) (Header Made from 2 Boards) | | | | |
	2 × 4	2 × 6	2 × 8	2 × 10	1 × 12
Roof and ceiling load	4'	4'6"	6'8"	8'10"	11'
1 floor, roof, and ceiling load	NAᵃ	4'	4'6"	6'8"	8'10"
2 floors, roof, and ceiling load	NA	NA	NA	4'6"	6'8"
Garages or nonbearing wall	6'	6'8"	8'10"	11'	13'

ᵃNA—not available

you will need the thickness of two studs (1½ inches + 1½ inches + ½ inch of plywood thickness = 3½-inch thickness for the header).

At the inside of each of the full-length king studs, another stud, called the trimmer stud, is added. The trimmer stud is cut to fit in the door opening from the bottom edge of the header down to the floor so that the header ends sit on the top ends of the pair of trimmer studs. The header and trimmer studs complete the rough opening for doors. For windows, the trimmer studs fit from the bottom edge of the header down to the top edge of the windowsill. The trimmer studs not only add strength to the opening, via the doubled studs, but also support the header above the opening, transferring the weight from above the header down to the sole plate at the floor.

For windows, a 2 × 4 sill is set at the point where the bottom of the window will sit in the rough opening, and the cripple studs are installed from the bottom edge of the sill down to the sole plate. Then trim-

mer studs are installed between the rough sill and header, completing the window rough opening.

 PLANNING POINTER

Installing Windows at Eye Level

For design uniformity, window headers are set at the standard height for door headers, that is, 6 feet 8 inches from the floor. But the height of the window and the windowsill can vary, and you should choose window sizes so that the window is at eye level for persons within average height ranges. Window height depends also on whether one is normally sitting or standing in a given room. To preserve privacy, shorter windows are used in bedrooms.

Finish Carpentry

Interior Doors

Interior doors once were wood slabs that had to be cut to fit the opening, then cut for hinge mortises and locks. Today, doors come in packages that include the prehung door, and frame and trim moldings. The hinges are in place, and the door and frame are bored to receive the lockset and strike plate.

Most modern interior doors are hollow core, rather than solid wood. Hollow-core doors consist of a framework of narrow wood strips, with cardboard filler webs glued to the interior of the door. The exposed surfaces of the doors are thin wood veneers. The reason for the move toward hollow-core doors is primarily for economy: solid wood or paneled doors are much more expensive than hollow-core doors.

Trim and Moldings

Trim and moldings are used to cover joints or cracks where surfaces or building materials meet, such as where the wall meets the ceiling or floor, or where a doorjamb meets plaster or wallboard.

Moldings can also be used for purely decorative purposes. For example, cove moldings are often installed for decorative purposes, where the walls meet the ceiling. Decorative moldings can also be used to enhance the entire interior. Chair rails are often installed in dining rooms or kitchens; they not only protect the wall from being damaged by chairs (thus the term *chair rail*), but also provide a decorative touch. Often, the homeowner will elect to use wallpaper on the upper portion of the wall above the chair rail, and easily cleaned, washable paint on the wall portion below the chair rail. Other uses for decorative trim include L-shaped wooden molding to protect the outside corners of a wall, and plate rails in the kitchen to dress up cabinet or other doors.

Installing Moldings and Trim

For installing moldings and trim, you will need a miter box and back saw, a coping or fine scroll saw, a carpenter's square, a sharp pencil, a quality rule, and a small hand plane. In most cases, cutting and mitering moldings or trim is a straightforward project. Moldings that will lie flat on the surface when installed are simply laid flat against the back, or fence, of the miter box and cut at the required angle.

Since few walls actually intersect at a true 90-degree angle, moldings that meet at inside corners are usually joined with a coped joint. This joint is not hard to make. The first piece of base trim is cut square and fit to the corner. Then the end of the joining piece of base trim is cut at a 45-degree angle. The 45-degree cut will expose the profile of the molding at the inside edge of the cut. Use a coping saw to cut along the profile to remove the wood at the end of the molding. By cutting the base trim in this way, the trim will always have a tight fit at the joint, even though the corner where the base meets may not be a true square, or 90-degree, corner.

For a crown or cove ceiling molding, also called a sprung cove, you must set the piece of molding in the miter box exactly as it will set against the wall/ceiling joint; that is, you must cut it at the angle it will form to the wall/ceiling. In this case, you will put the end of the cove molding in the miter box so that the surface that will set against the wall is on the table, or base, of the miter box, and the surface of the molding that will set against the ceiling is against the fence, or back, of the miter box. Then swing the saw to cut the corner at the appropriate angle and cut either a left or right miter as needed.

Most cove moldings can be cut by positioning an angled filler block in the miter box and resting the cove molding against the filler block. Cut one angle of the filler block at 38 degrees, the other angle at 52 degrees.

Stair Layout

There is a rhythm to climbing stairs. For safety, to avoid tripping, the steps in a staircase should be uniform, that is, all the steps should have the same dimensions for the height of the risers, which are the back of the stairs, and the width of the treads, which you actually step on. The stair treads also overhang the risers and this overhang is called a nosing. To help ensure stairs meet these requirements, building codes specify the overall minimum dimensions that are needed to make a safe stairway. Building codes generally allow a maximum riser height of 8¼ inches, a tread width of 9 inches, and a nosing of at least 1 inch (see figure 3.5).

To calculate measurements for a staircase, first figure the rise, which is the vertical distance between floors, and the run or horizontal distance between the beginning of the stair and where it will end. Divide the total rise in inches by 7, then round the

Figure 3-5   Stair Layout

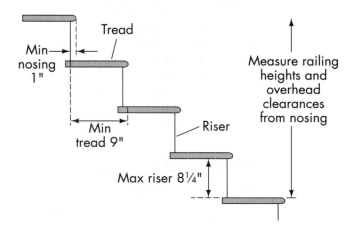

fraction to the nearest whole number. This will give you the number of risers. Divide the total rise in inches by the number of risers to determine the height of the risers. To figure the width of the treads, divide the total run in inches by the number of treads (one less than the number of risers). When planning a stair, if the height of the risers or

Building Safe Stairs

A general rule of thumb for stair layout is that the sum of the dimensions of one riser and one tread should fall between 17 inches and 18 inches.

A flight of stairs will be comfortable to use if two times the height of one riser, plus the width of one tread, is equal to 26 inches.

When repairing a damaged staircase, keep in mind that all the various stair components are available from many lumberyards or millwork outlets as stock mill items.

Hand railings for stairs should be 1⅝ inches in diameter, so the fingers can completely encircle the railing to maintain a firm grasp of the rail. Be sure the rail bracket screws penetrate through the wallboard and at least 1 inch into wall studs, so the rail will support the weight of a falling person.

threads exceeds the code maximums, go back and add or subtract one riser, then refigure your riser/tread numbers so they fall below the listed maximums.

Note also that safety and convenience dictate minimum dimensions for stairs. The minimum width of a staircase is 2 feet 8 inches; minimum overhead clearance is 6 feet 8 inches; hand railings should be installed 2 feet 8 inches to 3 feet above the top of the treads.

Hardwood Flooring

Hardwood flooring offers durability, easy maintenance, and good looks. Although oak accounts for 80 percent of all flooring installations, flooring is available in several other species, including maple, beech, and pecan.

Most hardwood flooring is tongue and groove, and is available in widths of 1 to $3\frac{1}{2}$ inches, and thicknesses of $\frac{5}{16}$ to $\frac{25}{32}$ inches. The thinner hardwoods are the most popular with homeowners for do-it-yourself or remodeling installations, because the thinner $\frac{5}{16}$-inch-thick flooring strips can be installed directly over existing floor finishes, such as linoleum or vinyl.

Some types of hardwood flooring must be nailed down, but many flooring manufacturers offer prefinished strip flooring that is either glued directly to the old flooring or laid over a thin foam layer, with flooring strips edge-glued to each other.

If you are remodeling your home, consider using prefinished strip flooring for your new floors. The prefinished strips are sanded and coated with a finish at the factory, so that the amateur installer avoids two difficult jobs: floor sanding and floor finishing.

Be sure the subfloor surface is clean and squeak free. Now is the time to renail old flooring and eliminate any squeaks before installing new flooring. If you will be using adhesives to fasten the flooring to the old floor, inspect the surface linoleum or vinyl to be sure it is securely stuck to the old plywood subfloor.

Follow the manufacturer's instructions regarding installation procedures and adhesives. If you nail the flooring strips down, rent a floor nailer to avoid damage and hammer marks to the flooring strips. Use a miter saw to cut the flooring strips to length. Stagger the end joints between flooring rows to avoid a continuous joint across the floor.

To avoid warped or buckled floors, allow a margin for expansion, usually about $\frac{1}{2}$ inch, between the new flooring and the walls. This margin or gap at the edges will permit the wood flooring to expand and contract as temperatures and humidity change with the seasons. Use base shoe molding to cover the gap between the new flooring and the base trim.

PLANNING POINTER

Protecting Hardwood Flooring

Hardwood flooring should be protected from moisture during the delivery and storage phases. Store hardwood only in a warm (70 degrees), dry area. Open flooring packages and let the flooring stand at room temperature to acclimate it to the temperature and humidity level in the house.

Cabinet Layout

When remodeling a kitchen, keep in mind the design principle known as the work triangle. This is the triangle formed as the cook works among the sink, refrigerator, and range. The point is to keep these distances limited, to avoid having to move about while preparing meals. The distance from sink to refrigerator should be 4 to 7 feet; from sink to range, 4 to 6 feet; and from range to refrigerator, from 4 to 9 feet. The minimum total of the three sides of the triangle is thus 12 feet, and the maximum is 22 feet. Ideally, the perimeter of the triangle should be between 15 and 20 feet.

Cabinets are designed to fit underneath a soffit or dropped ceiling. The soffit is usually dropped 12 inches from the ceiling, leaving a cabinet space of 84 inches, or 7 feet, from floor to soffit.

Prefinished cabinets are manufactured to fit in a 7-foot-high space. Typical base cabinets are built to a height of 36 inches; wall cabinets, to a height of 30 inches. This leaves an 18-inch space from the countertop to the wall cabinet. Often, range height

RULE OF THUMB

Planning Cabinet Space

Plan about 12 square feet of cupboard space for glassware and china, and an additional 6 square feet per family member for general storage in a kitchen.

is 36 inches, and a wall cabinet with a height of 18 inches is installed above the range. This provides a space of 30 inches from range top to wall cabinet. This 30-inch space permits for the installation of an exhaust hood under the wall cabinet and also allows space for the heat from the range to dissipate before it reaches the wall cabinet. The average refrigerator is about 64 inches high, so shorter wall cabinets, usually between 12 and 18 inches high, are used above refrigerators.

4 Plumbing

A generation ago residential plumbers used steel pipe for the water supply, and steel or cast iron pipe for the drain/waste/vent system. In those days plumbing projects required the use of such expensive and complicated tools as steel pipe cutters or power hacksaws, various sizes of pipe dies for threading pipe, and torches for handling the molten lead used for sealing the bell joints in drain pipe.

But complicated tools and materials were not the only barrier to do-it-yourself plumbers. In many areas, building codes forbade any but professional plumbers from doing the work, and this prohibition was further reinforced by tough union rules, which jealously guarded the plumbers' turf. The homeowner was restricted to the basic jobs of repairing a leaky faucet or using a plunger or snake to open a clogged drain. Major plumbing work was usually left to professionals.

Plumbing methods and materials changed drastically during the building boom following World War II. Steel water supply piping was replaced by copper piping that was easy to cut and to join with solder. The old lead-joint system of joining drain/waste/vent piping gave way to no-hub pipe and fittings.

Plastic pipe and fittings became popular during the 1970s. The plastic pipe was easy to work with, and could be joined via molded fittings and pipe cement or solvents. All these advances in materials and techniques made plumbing work more simple, and easing of restrictive codes further advanced the field of do-it-yourself plumbing. Today, most plumbing jobs are within the abilities of the homeowner, subject to local permit restrictions.

Building Codes and the Home Plumber

The plumbing in a house carries a supply of potable water (safe for human consumption) into the house, and carries human waste and household chemical wastes away to a sewage facility for treatment. (See figure 4.1 for a typical residential plumbing system.) Because of the potential dangers to public safety and health, plumbing work is subject to strict regulation and code control.

Figure 4-1 Residential Plumbing System

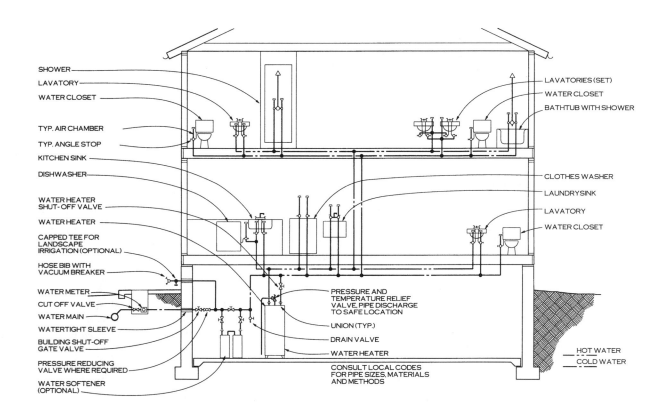

SHOWER
LAVATORY
WATER CLOSET

TYP. AIR CHAMBER
TYP. ANGLE STOP
KITCHEN SINK
DISHWASHER

WATER HEATER
SHUT-OFF VALVE
WATER HEATER
CAPPED TEE FOR
LANDSCAPE
IRRIGATION (OPTIONAL)

HOSE BIB WITH
VACUUM BREAKER

WATER METER
CUT OFF VALVE
WATER MAIN
WATERTIGHT SLEEVE
BUILDING SHUT-OFF
GATE VALVE
PRESSURE REDUCING
VALVE WHERE REQUIRED
WATER SOFTENER
(OPTIONAL)

LAVATORIES (SET)
WATER CLOSET
BATHTUB WITH SHOWER

CLOTHES WASHER
LAUNDRY SINK
LAVATORY
WATER CLOSET

PRESSURE AND
TEMPERATURE RELIEF
VALVE, PIPE DISCHARGE
TO SAFE LOCATION

UNION (TYP.)
DRAIN VALVE
WATER HEATER

CONSULT LOCAL CODES
FOR PIPE SIZES, MATERIALS
AND METHODS

HOT WATER
COLD WATER

Today, in most jurisdictions, the homeowner may do ordinary plumbing repairs without obtaining a permit. A permit is required for any major alteration to the plumbing system, such as remodeling or adding a bathroom, but the homeowner can still do the work subject to obtaining a permit and undergoing periodic inspections to ensure compliance with plumbing codes.

In some building jurisdictions, the code may require that major jobs be done only by a licensed plumber. Keep in mind that building codes vary from one city to another, so be sure to check with your own city's building department to learn the permit and code requirements in your area.

Plumbing Layout/Design

Because of rising demands on home water usage, such as lawn sprinkler systems, ¾- to 1-inch-diameter supply pipe is required from the street to the water meter in a new house. From the main water supply pipe, smaller ½-inch-diameter pipe is used to supply water to the plumbing fixtures (bath, sinks, toilets). Smaller ⅜-inch risers are used to connect the water supply pipe to faucets at sinks, or to the bathroom water closet (toilet tank).

In order to make installation of plumbing fixtures easier, manufacturers include the location of the drain and water supply lines on many of their fixtures. When plumbers install drains and water supply lines, they use these published roughing-in dimensions to locate the pipes. Many fixtures like bathtubs and toilets have pretty standard roughing-in dimensions. Figure 4.2 shows the location of the drain and water supply lines for standard plumbing fixtures.

Bathroom Layout: Piping and Fixtures

Modern houses are designed with more than one bathroom and with bathrooms of ever increasing size. Plenty of square footage is allocated to the modern bathroom, but that is not true in older homes. Finding space for bathroom expansion can be a challenge.

When laying out a bathroom, there are minimum requirements as to spacing and location of the fixtures to obtain a usable bathroom. The layouts in figure 4.3 show the minimum spacing between the bathroom walls and the various bath fixtures (tub to toilet, lavatory to toilet, etc.); for more generous sizes, increase the distance

Figure 4-2 Plumbing Supply Lines for a Bathroom

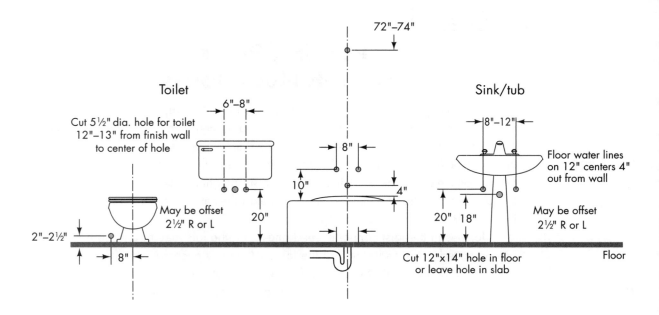

Figure 4-3 Bathroom Layouts

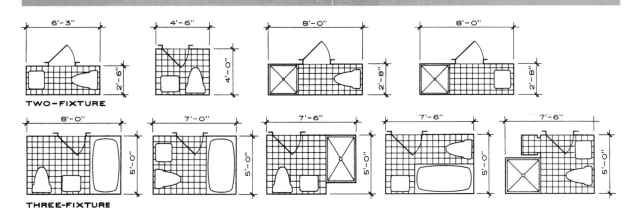

between the fixtures to make the bathroom more comfortable and accessible to all.

When designing a new bathroom, wheelchair accessibility should be considered a design priority. As the population ages, sooner or later most people will find a conventionally designed bathroom difficult or impossible to use. A bathroom designed with accessibility in mind will ensure that you will get the best return on your remodeling dollar, since you will be able to live in your house longer. Figure 4.4 shows the minimum bathroom clearances for wheelchair users and the location of grab bars.

RULE OF THUMB

Routing Pipes through Floor Joists

When installing plumbing, it is often necessary to route a pipe through a hole in a floor or ceiling joist. To avoid weakening the joist, there can be no holes through the joist that have a diameter more than one-fourth the joist depth. For example, there can be no hole larger than 2½ inches in diameter through a 2 × 10 joist. Holes that are on the centerline or midpoint of the joist's height can occur anywhere along the joist length; if the hole must be above or below the centerline, it cannot be placed in the center half of the joist along its length.

Notches in the joist cannot exceed one-fourth the joist depth, and no notches are permitted in the center half of the joist. When it is necessary to notch a joist, the code may require that you use a formed piece of metal, called a "reinforcing shoe" or guard plate, to reinforce the joist and preserve its strength.

Figure 4-4 Bathroom Accessibility

Lavatory Clearances

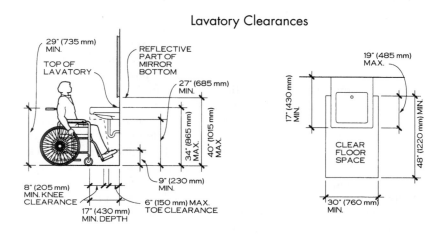

Location of Grab Bars

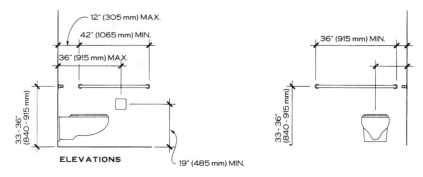

Pipes

Water Supply Pipes

Water supply pipes distribute water throughout a house. The three kinds of water pipe include steel, copper, and plastic.

In houses built before the 1950s, the water supply pipe was steel. Steel pipe is relatively long-lasting, but in time steel pipes rust through and must be replaced, either with new steel, copper, or plastic pipe.

Steel pipe. Steel pipe is threaded at the ends, and threaded connectors are used to join the pipes. During the threading process, steel is cut away in the threaded area, and the thickness of the pipe is thus reduced. Because the pipe will rust through the thinnest areas first, the pipe most often will fail and leak in the threaded area, near or inside the connector fitting.

If a steel pipe leaks, you can replace the failed section either with new steel pipe, or with copper or plastic pipe. Keep in mind, however, that if the steel pipe rusts through in one area, the entire system may be close to failure, and you may choose to replace the entire water supply system with copper pipe. A job this extensive usually calls for the services of a professional plumber.

Copper pipe. Copper pipe became the material of choice during the 1950s for water pipe. Copper is easy to work with, because it can be cut with a pipe cutter and joined with copper fittings, all soldered together using a propane torch. Copper pipe will not rust, and it resists corrosion and plugging from minerals.

Lead solder was used to join older copper piping, but has been forbidden for more than 10 years. In some cases, the lead may leach into the water and be ingested by the occupants. This risk is considered to be low in older houses, where the lead has leached out and the danger has diminished over the years. In most cases, the risk from lead can be minimized by letting the water run for a few minutes before using it for cooking or drinking. If your house is more than 10 years old and you have concerns about the lead content of your water, have the water tested by the public health department or by a private testing laboratory.

For your safety, if you have leftover lead solder in the shop, discard it and buy new lead-free solder for use in plumbing repairs.

Plastic pipe. Plastic plumbing includes polybutylene (PB), polyethylene (PE), chlorinated polyvinyl chloride (CPVC), polypropylene (PP), rubber-modified styrene (SR), acrylonitrile-butadiene-styrene (ABS), and polyvinyl chloride (PVC).

PB pipe is flexible plastic tubing used indoors as hot or cold water supply pipe. PB materials cannot be solvent-welded and are connected via mechanical fittings. An example is the Uncopper line by Genova Products, Inc.

PE pipe is a plastic pipe used only outdoors and below ground to carry cold water. This type of pipe is used for underground sprinkler systems.

CPVC pipe is a heat-resistant, rigid plastic pipe joined via solvent-welding and used for hot and cold water supply pipes.

PP pipe is a semirigid plastic pipe that resists heat and chemicals and so is a favorite for drain pipes. Polypropylene cannot be joined by solvents. Use mechanical (slip jam nut) couplings to join PP pipe.

Styrene (SR) pipe is a plastic piping used for drain pipe. It can be joined by solvent-welding.

ABS pipe is a plastic pipe used for drain/waste/vent systems. ABS pipe can be joined by solvent-welding.

PVC pipe is used for drain waste, vent, and lawn sprinkler systems and is joined by solvent-welding.

Drain/Waste/Vent Pipes

The drain/waste pipes carry water and household waste from the house to the main sewer pipe (see figure 4.5). Unlike water supply pipes, drain/waste pipes are not pressurized. Drain/waste pipes are installed with a slope of ¼ inch per foot, so that waste will flow by gravity through the pipe.

Maintaining the slope of the drain pipe is important. If the slope is less than ¼ inch per foot, the drain will be slow and solid waste may plug the drain. If the slope is more than ¼ inch per foot, the speed of the waste flow may be such that it will create a suction in the drain pipe and siphon the water out of the traps.

A properly installed and maintained drain/waste system should stay free of blockage. The fact is that drain/waste pipes may occasionally become plugged, either by accident or from negligence. For this reason, cleanout plugs are installed in drain/waste pipe at any point where the pipe changes direction, or at intervals of not more than 50 feet.

There is no such thing as an empty pipe. As the water or waste runs out of the pipe, air must flow in to avoid a vacuum. In order to provide this airflow into the pipe, vent pipes are installed above the drain/waste pipes. The main plumbing vent pipes rise in the kitchen wall, not more than 36 inches from the trap seal, and in the bathroom above the waste pipe for the toilet stool.

This main toilet vent is usually 3 or 4 inches in diameter. (Check your local code.)

Other fixtures that must be vented include the bathroom vanity sink drain, the bathtub drain, and laundry drains. Rather than run separate vent pipes through the roof for each of these drains, smaller pipes, called branch vents or revents, are used to connect the drain/waste pipes to the main bathroom vent. Your local code may set limits on how long the branch vent may be, and on how many fixtures may be vented to the main vent.

Figure 4-5   Drain/Waste/Vent System

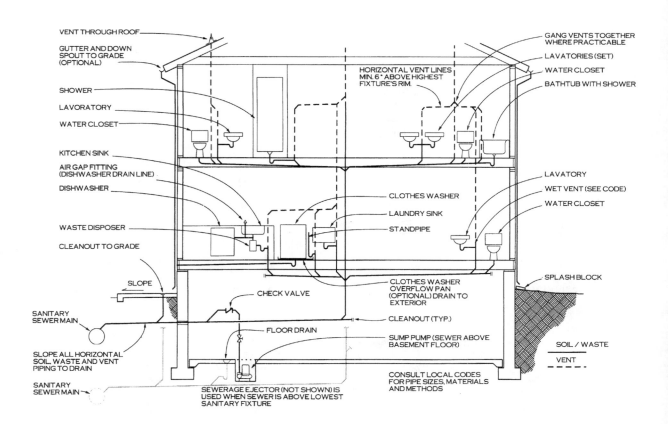

Fittings

The term *fittings* refers to the connectors used to join two or more lengths of pipe (see figure 4.6). Fittings are available to join pipes end to end; to change direction (e.g., a 45- or 90-degree elbow); to change sizes (e.g., a ¾-by-¾-by-½-inch tee that connects a ½-inch branch line to a ¾-inch supply line); or as adapters to join different types of pipe like copper to steel. If you draw a diagram of the pipe layout you plan, the dealer can help you choose the fittings needed. Many home centers sell only pack-aged fittings, but some full-service hardware stores will still cut and thread steel or galvanized pipe to order. Measure the pipe carefully to be sure you have it cut to the right length.

The fittings for either steel or copper pipe are universal and will fit any same-size steel or copper pipe. But fittings for plastic pipe (see figure 4.7) are not universal in size between manufacturers, and plastic-solvent adhesives will not fill a loose-fitting pipe connection. Always buy all plastic pipe and fittings from the same dealer to ensure you are getting the same brand.

Figure 4-6 Water Supply Fittings

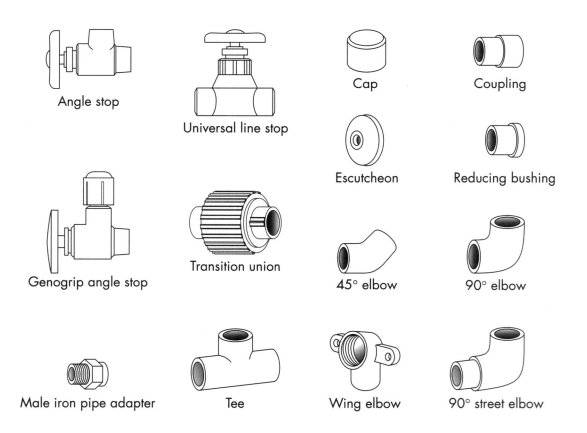

Angle stop

Universal line stop

Cap

Coupling

Genogrip angle stop

Transition union

Escutcheon

Reducing bushing

45° elbow

90° elbow

Male iron pipe adapter

Tee

Wing elbow

90° street elbow

Figure 4-7 Plastic Fittings

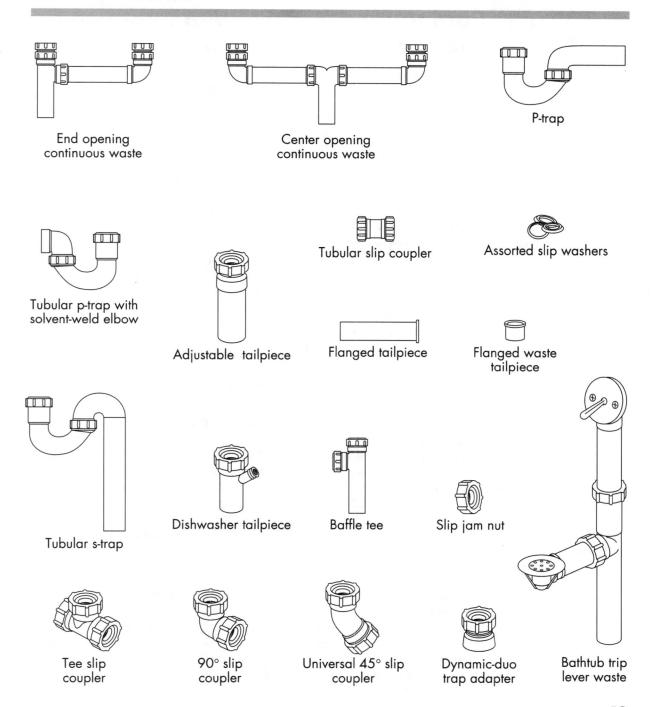

End opening
continuous waste

Center opening
continuous waste

P-trap

Tubular p-trap with
solvent-weld elbow

Adjustable tailpiece

Tubular slip coupler

Assorted slip washers

Flanged tailpiece

Flanged waste
tailpiece

Tubular s-trap

Dishwasher tailpiece

Baffle tee

Slip jam nut

Tee slip
coupler

90° slip
coupler

Universal 45° slip
coupler

Dynamic-duo
trap adapter

Bathtub trip
lever waste

59

RULE OF THUMB
Cutting Pipe Accurately

When cutting pipe, always allow for the space taken up by the fitting, and the distance the pipe slips or screws into the hub of the fitting. This allowance is called "makeup."

Solvents/Adhesives for Plastic Pipe Fittings

Plastic pipes that can be solvent-welded include PVC, CPVC, ABS, and styrene or SR. To ensure foolproof joints, always buy pipe and fittings made by the same manufacturer. The pipe should have some resistance when pushed into the fitting, and should not fall off when the pipe and fitting are held upside down (inverted).

Solvent cement combines plastic filler with a solvent. When the solvent evaporates, the dissolved plastic literally welds the pipe and fitting together. This material will not fill a wide gap between pipe and fitting, so it is important to be sure that the pipe and fitting will mate properly.

All Purpose Cement is an all-purpose solvent cement that can be used with any of the plastic pipe materials mentioned above.

To ensure a nonleaking joint, it is important to cut pipe ends square so they will seat properly in the fittings. Use a tubing cutter or miter box to be sure that all pipe ends are cut square. Use a sharp knife to clean away any burrs from the cut ends. Next, clean all mating surfaces of the pipe and fittings with Novaclean cleaner/primer. When the pipe is clean and dry, coat all mating surfaces with All Purpose Cement.

Plumbing Maintenance and Repair

Pipe Repairs

The following are the most common plumbing problems, along with their appropriate repairs.

Sweating Pipes

Water pipes sweat when humid air contacts cold water pipes, which causes the water vapor or humidity to condense and form water droplets. This condensation or sweating can cause wet floors or cabinets wherever the sweating occurs: in laundry rooms, bathrooms, or under vanity or kitchen sinks.

To stop pipes from sweating, insulate water pipes. Preformed foam plastic pipe insulation is available at home centers. Just slip the insulation over the pipe and use duct tape to secure it in place.

Noisy Pipes

Water pipes may become noisy due to one of two reasons. First, expansion and contraction of the pipes (especially hot water pipes) may occur if the pipe hangers hold the pipes too tightly against a joist. This noise can often be cured by loosening the pipe hangers so the pipe can expand and contract without creating noise.

Second, water hammer occurs when water flowing at high velocity through a pipe is suddenly stopped by a closed valve or faucet. Because water cannot be compressed, a hammering noise results when the water velocity is halted. Water hammer is particularly common at washing-machine faucets, because the valve in the washer snaps shut when the machine has filled.

Figure 4-8 How a Water Hammer Muffler Works

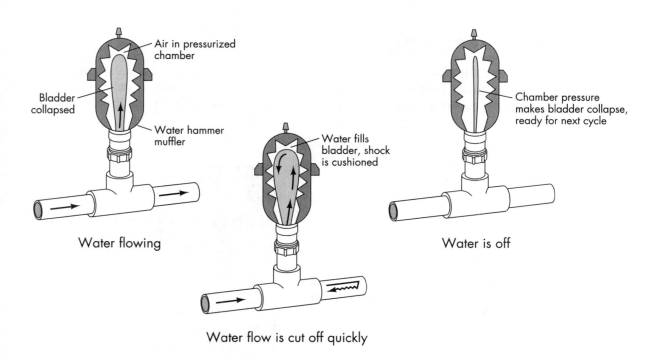

Air in pressurized chamber

Bladder collapsed

Water hammer muffler

Water flowing

Water fills bladder, shock is cushioned

Water flow is cut off quickly

Chamber pressure makes bladder collapse, ready for next cycle

Water is off

To eliminate water hammer, you must install water hammer mufflers or air chambers at the faucets where water hammer is heard. When the water flow stops suddenly, the air in the mufflers will compress, reducing the impact and eliminating the noise, or hammer (see figure 4.8). Install the water hammer mufflers as close as possible to the noisy faucet.

Types of water hammer mufflers include copper tubes that can be soldered into the pipes near the faucets, and plastic air chambers that can be installed in the water pipes. The easy solution to water hammer at a washing machine is to install copper mufflers that can be screwed onto the water faucets between the faucet and the washer hose. Look for these simple screw-on mufflers at your home center.

Frozen Pipes

To avoid problems with frozen water pipes, avoid routing pipes through unheated spaces in a garage or crawl space. If a pipe must pass through such space, or be exposed to freezing temperatures in an exterior wall, insulate the pipe carefully to prevent freezing. Electric pipe heaters are available that can be wrapped around the pipe to keep it from freezing.

RULE OF THUMB

Preventing Pipes from Freezing

To prevent water supply pipes from freezing before they enter the house, always install water pipes below the frost line—the depth at which the ground freezes, usually between 2 and 4 feet deep. (Check with your area's building department for your frost line.) If your water supply pipe freezes occasionally in very cold weather, let the water run at a trickle at any sink or tub inside the house. The flowing water will not freeze.

If a pipe does freeze, do not use a propane torch to thaw it. Too much heat may melt the ice in the pipe and turn it to steam, which can explode in confined spaces. Instead, use a blow-dryer and, with the faucet open, play the dryer over the length of the pipe to thaw the entire pipe at once, rather than overheat one small spot with concentrated heat.

Faucet Repairs

The most common plumbing job may be repairing a leaky faucet. To repair any faucet, you must know the type of faucet you have, and the name of the manufacturer. Three basic types of faucets are available. They are the compression (stem-and-seat) faucet, the ball-type faucet, and the cartridge (disk or sleeve) faucet. Repair procedure for each type of faucet is simple, but you must determine which type you have and be very careful that you obtain the right repair parts for your particular faucet.

Fixing Compression (Stem-and-Seat) Faucets

A common type of faucet is the compression, or stem-and-seat, faucet. The compression faucet is commonly used on double-handle sinks or bathtubs. These compression faucets have a neoprene washer at the base of the stem. Water control is achieved when the faucet washer is compressed against the faucet valve seat to seal it.

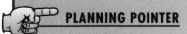

PLANNING POINTER

Buying Faucet Repair Parts

Repair parts vary not only by type of faucet but by manufacturer, so learn the name of the manufacturer before shopping for repair parts.

If the manufacturer's name is not visible on either the index cap or the faucet body, take the defective part to the plumbing dealer and ask for a matching replacement. When buying washers and O-rings to repair a leaky compression faucet, it is best to buy an assortment pack, or take the faucet stem and washer along to the plumbing shop. Each model of faucet takes a different washer and O-ring. The same advice applies for kits to replace ball-type or cartridge (sleeve or disk) faucets. Remember always to shut off the water supply before working on faucets.

First, shut off the faucet's water supply. To disassemble the faucet, remove the screw holding the handle to the faucet. Carefully lift or pry upward to remove the faucet handle. Remove the retaining nut or packing nut, and remove the stem from the faucet body. A stem screw secures the washer on the end of the stem. Remove the stem screw and the old washer. Select a matching replacement washer and install the new washer.

The stem O-ring prevents water from leaking upward past the stem. Always replace the O-ring when you replace the faucet washer. Use a razor knife to cut the old O-ring off the stem. Apply heat-proof plumber's grease on all moving parts. Replace the O-ring.

Note that on some older faucets the O-ring may be omitted and there will be a packing string or packing washer on the stem instead. In this case, replace the packing string or washer. Reassemble the faucet and check for leaks.

Fixing wall-mounted compression faucets. Your bathtub may be served by wall-mounted compression faucets. Replacing the washers on these faucets is the same as above, but you will need an extension socket and ratchet wrench to remove the bonnet nut on these faucets. Most plumbing supply stores will have a set of "loaner" extension sockets that you may borrow for faucet repairs.

Fixing Ball-Type Faucets

Shut off the faucet's water supply. Use an Allen wrench (often supplied in repair kits) to loosen the handle set screw. Remove the faucet handle to gain access to the adjusting ring. Use the wrench to tighten the adjusting ring. Put the faucet handle back on and turn on the water to check the faucet. If the faucet still leaks, you will have to remove and replace the O-ring and seats and springs.

Fixing Cartridge (Disk or Sleeve) Faucets

To repair a leaky disk-type cartridge faucet, first shut off the faucet's water supply. Remove the index cap and handle set screw, then lift up the faucet handle. To remove the handle insert, use an Allen wrench to loosen the set screw, then unscrew the dome cap and lift it off. Next, remove the cartridge mounting screws and lift out the cartridge. Now insert the new cartridge and replace the mounting screws. Screw on the dome cap and reinstall the handle insert, handle, and index cap.

To fix a leaky sleeve-type cartridge faucet, pry up the index cap and remove the lever. You must lift the lever to the maximum position to free the inner lever from the lip on the retaining nut. Use adjustable pliers to remove the retaining nut and the grooved collar if one is present. Cut off and replace the O-rings, then reinstall the spout and retaining nut. Again holding the lever in its uppermost position, slip the flat edge of the inner lever over the rim of the retaining nut. Install the handle set screw and the index cap.

Avoiding Plugged Drains

A household drain is designed to carry away only waste water, human waste, and, if the home is equipped with a garbage disposal, kitchen waste. A household drain/waste system that is properly used and maintained should have few if any plugs or stoppages. The key is never to use the drain/waste system as a disposal for other materials. Most plugged drains are caused by occupant error, usually by carelessly introducing into the drain materials that were never intended for sewer disposal. Following is foolproof advice for avoiding plugged drains.

Foolproof Advice
for Avoiding Plugged Drains

- Do not dispose of plaster or other patching materials in the sewer or drain. Plaster and other patching materials will harden in the drain pipes and clog them. Dump the patch materials on newspapers and dispose of them in the trash.

- Do not dump waste such as coffee grounds down drains or toilets. If you have a garbage disposal unit, follow the manufacturer's directions regarding materials that can be run through the disposal unit.

- Never put into the toilet anything you didn't eat. Toilet tissue is made to degrade in the sewer, but other paper materials (including tissues and paper towels) are not. Put bathroom-generated waste into a wastebasket, not into the toilet.

- When using the garbage disposal, always use water to flush down food particles and carry them into the main (large) drain pipes.

- Always use a hair strainer over the drain in a bathtub, because most bathtub drains are plugged by soap bits and human hair.

- Use a drain filter that fits over the end of the washer drain hose to prevent laundry lint from clogging the drains.

- Do not wait until you have a sewer drain backup and a flooded basement to clean the main drain. Drain or sewer cleaning should be planned as routine maintenance, not as an emergency procedure to be done after your house or basement is already flooded. Your plumber can tell you how often the drains should be cleaned.

Replacing a Sink Trap

Hidden away under most plumbing fixtures, installed in the drain pipe, there will be a U- or J-shaped pipe called a trap. Traps perform two functions. The first and most important function of a plumbing trap is to trap and retain water in the lower bend. The trapped water blocks the drain pipe so that odors and sewer gases cannot flow back through the drain and into the house. The trap also provides a first line of defense against plugged drains by catching any materials that might form a plug farther down the drain, where it would be more difficult to reach and remove. Traps are easily removed for access to the piping and plug if a plugged drain does occur.

Because metal drain traps are always full of water, they are constantly exposed to rust and corrosion, and they are thus subject to frequent rust-out and replace-

Cleaning Sewer Drains

Sewer drains that serve a house with a large family, or drain pipes that are subject to root entry from a nearby tree, require frequent cleaning. Root-free sewer pipes that serve a small family may require little or no cleaning.

ment. Today's plastic traps are resistant to corrosion and failure, and may literally last forever, but in time the metal traps will fail and require replacement.

To replace a trap, the only tool you'll need is a pipe wrench or a pair of adjustable-jaw pliers. Some metal and plastic traps may be interchangeable, but we've found that to ensure a good match and avoid frustration, it is usually easier to replace a metal trap with metal or a plastic trap with plastic.

Maintaining the Water Heater

Water heaters perform heavy-duty service for years, but a few simple steps can extend the heater's life and keep it operating properly.

The first indication of water heater failure may be no hot water, or you may see water running out from under the heater—sure signs that the heater needs attention or replacement.

If your water supply has a high mineral content, minerals or sediment that accumulate at the bottom of the water heater can insulate the water from the burner's heat,

extending recovery time for the water heater and wasting energy. Place a plastic pail under the drain valve near the bottom of your water heater, or attach a garden hose to it, then turn on the water and let it run until it is clear and free of any sediment.

On the side of the water heater, you will find a water tube and a small valve. This is the pressure release valve—designed to open if the heater malfunctions to prevent dangerous steam buildup or possibly an explosion. To be sure the valve is operable, place a catch pail under the end of the relief tube and hold up the relief valve until water runs out of the tube.

To test the vent draft, light a match and hold it under the hood on the exhaust vent at the top of the water heater. If the vent is open, the flame from the match should flicker upward. The hood and vent pipe are secured in place into the larger furnace duct pipe by one or two small sheet-metal screws. If your heater is several years old, remove the vent pipe and clean it, then reinstall it in place.

Remember, to ensure that no one will be burned or scalded by hot water from the heater, keep your thermostat set at or below 110 degrees.

Repairing Washing Machine Hose Valves

In the laundry area, a pair of washing machine hose valves (compression faucets) control the water supply to the automatic washing machine. These faucets are used to shut off the water supply to the washing machine during repair or replacement of the machine. Always shut off the hose valves when the washing machine is not in use. If the hose valves are left open full time, the constant water pressure may cause the rubber hoses to break and flood the laundry area. To ensure that the hoses do not rupture, replace them every 3 to 5 years.

In time the hose valves will develop leaks, either through the faucet spout or around the stem. Because the heat from hot water will deteriorate the stem washer faster than cold water will, the hot water faucet usually will begin to leak first. Because you have to turn off the water supply at the meter to repair the faucets, it is best to repair both hot and cold water hose valves at the same time, while you have the proper tools and replacement washers handy and the water is turned off.

To take the hose valves apart, use a pipe wrench or adjustable pliers to turn and loosen the nut on the top of the faucet, and pull out the faucet stem and replace the washer.

At the top of the stem, under the retaining nut, you will find an O-ring or graphite-coated stem packing. Remove the old packing and replace it with new packing. To test for leaks, reassemble the faucet and turn the water on.

Repairing Toilets

Repairing a Running Toilet

When the toilet flush lever is pushed, the flapper ball is lifted to open the hole, and the water flows into the toilet bowl. After the toilet is flushed, the flapper ball drops back to shut off water flow into the toilet bowl. The toilet tank refills with water via the toilet flush valve. A float attached to the flush valve rises as the tank fills, and when the water reaches the tank's full mark, the rising float shuts off a water valve. The toilet tank is now full, waiting for the next flush action.

The flapper ball is made of rubber or plastic. If the flapper ball becomes worn, it may not seal the tank completely, and some water may leak past the flapper ball. This not only wastes water, but is a nuisance because of the continuous sound of water running into the toilet bowl.

The flapper ball blocks the flow of water into the valve seat. On older ball cock/float rod toilets, a tank ball blocks the water flow. If you have a toilet that runs continuously, first check the chain that connects the flush handle to the tank or flapper ball. If the chain is too short, it may not let the flapper ball close the hole completely. The cure is to lengthen the chain so the hole can close completely. If the chain seems to be set at the proper length but there is excess chain hanging below the flapper ball, the extra chain may be dropping between the flapper ball and the flush valve seat. The cure for this problem is to use wire-cutting pliers to cut off the excess chain.

If the flapper ball appears to be worn or cracked, remove and discard the old flapper ball, then install a new one. On older ball cock/float rod toilets, simply unscrew the float ball from the threaded lift wire, and then screw the new float ball onto the lift wire.

Replacing a Toilet Flush Valve

New toilet flush valves are inexpensive, made of noncorroding plastic, and have a float cup that encircles the ball cock shank. This eliminates the old float rod and float cup. If the old ball cock is not working well, replace it with the $10 plastic model.

To replace the flush valve, first shut off the water supply at the valve under the toilet tank.

Flush the toilet to empty both the toilet tank and stool. Use a large sponge to remove any water remaining in the toilet tank. Place a pail under the water supply pipe to the toilet tank. Use an adjustable wrench to loosen the water supply pipe and the retaining nut that holds the flush valve to the toilet tank. Remove the old flush valve or ball cock hardware and dispose of it.

Set the new flush valve into the hole in the toilet tank. Secure it to the toilet tank with the plastic retaining nut and tighten. To adjust the water level in the toilet tank, pinch the spring clip on the pull rod and adjust the float cup position on the ball cock shank (upward for a higher water level, downward for lower).

Curing a Lazy Toilet Drain

The toilet may flush properly one time, and on the next flush the water may rise to the rim of the toilet bowl and go down in a lazy spiral. Use a toilet plunger or snake to check for a partial toilet drain plug.

If the toilet consistently flushes in the same lazy way, the problem may be mineral buildup inside the toilet bowl. When the toilet is flushed, the water runs into concealed channels within the stool or bowl. Water runs out from holes under the rim of the stool to rinse the sides of the toilet bowl. Also, look at the front edge of the bowl and you will see a small entry hole. This hole may become plugged with minerals from the water, so that the water flow is restricted and the toilet flushes with a lazy drain. Reach into the toilet bowl and run a finger into the hole at the bowl front. If there is significant mineral buildup, you will be able to feel the blockage.

To cure a mineral buildup problem, use a small screwdriver or chisel to chip away the buildup. After scraping the perimeter of the hole, flush the toilet to see if the water flow has been restored. If so, your problem is solved. If not, there are drain cleaners available to clean out the channels inside the stool, far up where you cannot reach with a scraper. To clean out the interior water channels of a toilet, you must pour the cleaner into the overflow pipe in the toilet tank. The chemical cleaners often contain acid, so read the label and handle such products with care. Describe the problem to your hardware/plumbing supply dealer. He will be able to direct you to a good mineral removal product. One such mineral removal product is SAN-TEEN, made by San-Teen Products, 1321 Seventh Street South, Hopkins, MN 55423. If this product is not available in your area, check with your dealer.

Maintaining a Septic System

If your house is beyond the reach of municipal utility services, you may have your own sewer or septic system. In a septic system, household waste generated by the bathroom, kitchen, and laundry—99 percent of which is water—runs through a drain/waste pipe into a septic tank. Nonliquids such as human waste and paper settle out in the tank. As the solids settle out, liquids in the tank may flow into a second tank, or into a second compartment within a single tank. When settling has occurred, the liquid sewage flows into a series of seepage pipes. These pipes have holes that permit the liquids to flow out and be absorbed into the soil or a gravel seepage bed. This is sometimes called a "leach field" because the liquids "leach" or are absorbed into the soil. Light or sandy soils will easily disperse the liquid seepage. Heavy or clay soils cannot easily absorb the liquids, so a trench may be dug and then filled with gravel to form a leach field. The trench is then backfilled with topsoil and seeded or sodded over.

The size of the septic tank will vary according to the size of the house. A two-bedroom house will have a minimum 750-gallon septic tank; a three-bedroom house, a 1,000-gallon septic tank; and a four-bedroom house, a 1,200-gallon septic tank. When buying a house, check the size of the septic system to ensure that the system is large enough to handle your family's needs. The size of the septic tank should be recorded at the local building department.

Foolproof Advice
for Maintaining Your
Septic System

- Repair any leaky faucets or running toilets to reduce the amount of waste that enters your septic system. Install water-saving devices, such as low-flow shower heads and low-volume toilets, to reduce water usage.

- Don't use your toilet as a waste disposal system. Only toilet tissue should be flushed down the toilet. Dispose of other waste paper, such as facial tissue and soap wrappers, in a wastebasket.

- Avoid using chemicals advertised to increase bacterial action in the septic system. Bacterial action will naturally occur in a properly built septic system. Putting the wrong chemicals in the septic system may kill the natural bacterial action that must be present to break down solids.

- Avoid using cleaning or bleaching chemicals that might interfere with natural bacterial action in the system.

- Don't neglect pumping the septic tank until problems appear. In an overloaded septic system, the solids enter and plug up the seepage pipes and ruin the leach field.

A properly built and maintained septic system will last for years. Natural bacterial action will break down the solids into sludge that can be pumped out by septic service companies. Waste liquids will seep into the soil in the leach field, and sewer gases will pass up the septic drain pipe and exhaust through the sewer vent system. Depending on how much waste your household generates, have your septic tank pumped every 3 to 5 years.

5 Electrical

Electricians are among the highest paid of the building tradesmen, so as a homeowner you can make significant savings by doing your own electrical repairs. With a little knowledge the average person can install new fixtures; replace electrical devices such as outlets, circuit breakers, and switches; repair lamps, doorbells, and chimes; and add extra outlets or outdoor lighting. For more complicated projects, such as wiring a new addition or bringing a home's electrical system up to current code, you may wish to do further study or hire a professional.

Electrical Wiring Permits/ Inspection

Electrical wiring permits are not needed for replacing electrical devices, such as switches or outlets, or lighting fixtures. Depending on local codes, more extensive electrical projects may require taking out a permit for the work and having the work inspected by the electrical inspector of your local building department. If you are in doubt whether your particular project requires a permit, check with your city's building department before starting the job.

Most electrical inspections require two trips: one at the rough-in stage to inspect the way the wires and boxes are installed, and a final inspection when the devices are installed after drywall and painting are completed. Do not cover up wiring circuits or boxes with wallboard or other finish before the rough-in inspection has been done, because the inspector may require you to remove the wallboard so that he can visually inspect the wiring.

Residential Wiring

How Power Gets to the Meter

Electrical power is generated at a power plant and sent via high-voltage cables to distribution stations. From these stations the power travels to local areas, where transformers step the voltage down for delivery to individual buildings.

RULE OF THUMB

The *National Electrical Code*

Before beginning your project, visit your bookstore or library for a copy of the *National Electrical Code (NEC)*. This book provides the basic guidelines for code-approved installation techniques, and its purpose is to protect the community against fires and serious shocks that can result from improper wiring. Be aware that local codes may vary in some respects from the national code, so if you have any questions, consult your local electrical inspector.

Two-Wire/Three-Wire Services

You can check how much electrical capacity your home has by counting the wires leading into the building. Older homes may have only two wires, one black (hot) and the other white (neutral). Such homes may have only 3600 watts of power available (watts = volts × amps). This was sufficient for the lower power demands of older homes, but is inadequate for today's power demands.

Modern homes have three entry wires. One wire is black (hot), the second wire is red (hot), and the third wire is white (neutral). You may pair black and white wires, or red and white wires, to provide a 120-volt circuit. Pairing a black wire with a red wire will give you a 240-volt circuit. This higher voltage is necessary to power large appliances and tools, such as electric ranges, water heaters, air conditioners, and welders.

The power demands of modern households require that the home be brought up to today's standards. Today's three-wire systems provide 100- or 150-amp service, and some have even greater capacity.

Service Entrance Panel

Where the electrical wires leave the voltmeter and enter the home, the power wires are attached to a service entrance panel. The white (neutral) wire is grounded at this point to the water supply pipe, where the pipe enters the home ahead of the water meter. This grounding protects the service and electrical equipment from damage.

Depending on the age of your home, there may be one of three types of service entrance panels. You should be aware of which type you have, and check how to shut off power to the entire home or the circuit you are working on.

The type of panel in older homes may be a fuse box. To cut power to the entire home, you must pull out the cartridge fuses marked MAIN. Appliances that use high current, such as clothes dryers or ranges, may have separate pullout fuse cartridges. Individual circuits are shut off by unscrewing the proper fuse. (See figure 5.1.)

RULE OF THUMB

Turning Off Power

Always remember to turn off the power source at the service entrance panel when you're doing any electrical project.

The second type of panel is the lever box. To shut off power to the entire house, you must pull down the lever and remove the main fuses, then unscrew the proper fuse to cut power to individual circuits. Having to pull the lever or handle protects you when you must change a fuse.

The latest type of panel is the circuit breaker panel. Power is shut off to the entire home by switching the breaker marked MAIN to the off position. Individual circuits are marked with stickers noting which circuit each breaker controls. (See figure 5.1.)

Fuses

A fuse is a common protective device most often used in older homes. Lighting circuits are usually protected by a 15-amp fuse; kitchen circuits, where several small appliances may be used, have 20-amp fuses; and large appliances may have fuses of 30 amps or more. Also, cartridge fuses of larger capacity are used for main power fuses.

Fuses let electricity flow through a metal strip. If the circuit is overloaded by too high an amperage flow, the metal strip will melt and break the circuit. Fuses may blow if the circuit is overloaded, if the circuit experiences a short circuit where a live wire contacts a neutral or ground wire, or if a fuse is loose in a socket.

The older type plug fuses are screwed into sockets in the service entrance panel. They are now outlawed because a 20- or 30-amp fuse would often be screwed into a 15-amp socket, causing a serious overload that could lead to a fire hazard.

Types of fuses used today include plug fuses (now banned by the *National Electrical Code* for new homes), nontamperable (Type S) fuses, time-delay fuses, and cartridge fuses.

Type S fuses have nonremovable adapters for each fuse size, so one cannot substitute a larger fuse for a smaller one.

Figure 5-1 Fuse Box and Circuit Breaker Panel

Fuse Box with Fused Main Disconnects

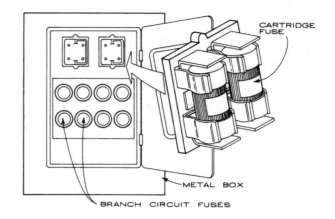

Circuit Breaker Panel

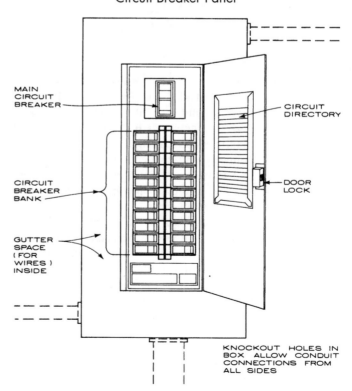

Checking Fuses

Plug fuses have a glass window in the top, so you can see the metal fuse strip. If the fuse has blown but the glass is clear, suspect an overloaded circuit and unplug some appliances on the circuit to reduce the load; if the glass is discolored on a blown fuse, suspect a short in the system or an appliance.

Some tools and appliances require a larger surge of power for starting, then draw fewer amps when running (e.g., air compressors). Time-delay fuses are designed so they will not blow under this temporary startup power load.

Making Electrical Repairs

Tools Needed

Tools needed for electrical repairs include a variety of slot and Phillips-head screwdrivers, long-nose pliers, a hammer, nails, and an electric drill. You will also need plastic electrical tape and twist-on wire connectors to connect wires together. Note that twist-on connectors come in several sizes, according to how many wires will be connected together.

If you plan to do more complicated projects, you will also need a combination tool that can crimp wire connectors and strip insulation from wires, and has a built-in wire gauge in the handle. A cable slitter that slits cable insulation without nicking the wires is a useful tool for new work or adding extra outlets. If you will be drilling through wood, you'll need a ¾-inch spade bit; if drilling through masonry, buy a ¾-inch carbide masonry bit. Bit extension

attachments are needed for drilling through multiple joists or beams. Buy a fish tape that can be used for pulling electrical wires through enclosed spaces, such as cavities between wall and floor framing.

For troubleshooting and safety, buy a circuit tester for testing whether there is power to a switch or outlet, and whether a circuit is grounded. The circuit tester has a neon lightbulb that lights if current is present.

Also buy a continuity tester. The battery-powered continuity tester tests for continuity in circuits or electrical devices when the current is shut off.

Continuity

To avoid problems in a 120-volt circuit, be aware of the term *continuity*. Continuity refers to the fact that throughout the electrical system, all black wires connect only to brass screw terminals or to other black wires; all white wires connect only to white (silver) screw terminals or other white wires; and green or bare ground wires connect only to ground screws or to other ground wires. Sometimes you may see a white wire used as a black wire, for example, at a three-way switch. But in this case the ends of the white wire will be taped or painted black to indicate it is being used as a black wire.

Many extension cords and appliance or power tool cords are polarized, meaning they have a neutral (white) wire plug prong that is wider than the hot (black) wire plug prong. Polarized outlets also have wider slots on the neutral side than on the hot side. Polarized plugs ensure that continuity is maintained through the extension cord or appliance or power tool cord. On flat lamp cords, the plastic insulation will be ribbed on the neutral or white wire side, so you should connect the wire from the ribbed side to the silver screw and the wire from the smooth side to the brass screw of the lamp switch. Keep in mind that only black (hot) wires are switched; by maintaining polarity, you ensure that when the switch is off there is no power to the lamp socket.

On tools or power cords that have a third (grounding) plug prong, the polarity and continuity are ensured when you insert the round ground prong into the electrical outlet.

Making Wiring Connections

Some electrical devices have push-in connections, so you merely have to strip the insulation off the end of the wire (there is a gauge on the device to indicate how much insulation to strip) and push the bare wire into the wire slot on the device. To be sure the wire snaps securely into place in the slot, pull gently on the wire.

If you are working with devices that have wire-binding screws, strip away enough insulation so you can wrap the wire in a loop at least two-thirds of the way around the screw. Wrap the wire around the screw in a clockwise direction, so the wire will be wound tighter as you tighten the screw. When you tighten the screw, give it an extra turn to be sure the wire is secured fast to the screw.

When attaching two wire ends together, strip about ¾ inch of insulation from the wires. Hold the two wire ends together so they are parallel and the ends are even. Use a plastic twist-on connector to secure the wires together. Just twist the connector clockwise until it is tight, then tug gently on each wire to be sure it is tightly secured. Do not use plastic electrical tape with twist-on connectors.

 PLANNING POINTER

Shopping for Proper Electrical Materials

- When shopping for electrical materials, always look for the Underwriters' Laboratories Inc. (UL) mark. This mark indicates that the device is approved by the laboratories and that it will provide safe and dependable service. You will find the UL marking on such devices as switches, outlets, connectors, extension cords, and wall plates.

- If you have aluminum wiring in your home, be sure to use wiring devices that are approved for use with either aluminum or copper wiring. These devices include receptacles, switches, and lamp holders. Such devices are marked CO/ALR or AL-CU. Also, when working with aluminum wiring, use an oxide-inhibiting compound to prevent corrosion at all joints.

- Check the instructions provided with the device. Some manufacturers offer more complete installation advice than others. The instructions may provide all the guidance you will need to finish your project.

Buy twist-on connectors in various sizes to be sure you have the right connector on hand. Use the smaller connectors to connect two wires together, larger connectors to connect three or more wires.

Using Home Plans

If you still have the original plans for your home, study the plans before beginning your wiring project. The electrical plans will show the original wiring system, and may be a useful guide to circuits, types of receptacles and lights installed, and other such information. The plans are an especially useful guide if you are planning a major addition to or renovation of the home.

The original electrical plan can be compared against the existing fixtures—switches, receptacles, and lights. Differences may mean remodeling has taken place. But remember, this plan is only useful as a guide because many times modifications are made during construction and not marked on the original plan. Figure 5.2 shows the meanings of various electrical symbols.

Figure 5-2 Electrical Symbols Used on Home Plans

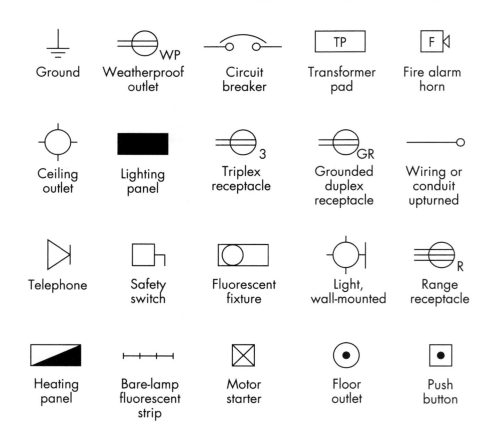

Updating Electrical Systems

If you have an older system, consider having an electrician bring it up to today's code, which has added power and safety features. These features will provide greater safety and convenience, add to the resale value of the home, and may even result in a lower insurance bill. Some insurance companies will discount home insurance premiums by 10 percent or more if you bring them a code certificate. Check with your insurance agent for details.

Branch Circuits

At the fuse or breaker panel, the power to the home is divided into several branch circuits before being routed through the home. By dividing the current into multiple circuits, lower-amp fuses or breakers can be used on each circuit. This permits you to shut down one circuit to work on it without interrupting current to the entire building. Major appliances are often connected to their own separate circuits, so that you can turn off the current to that circuit and still have power to the rest of the home. Having branch circuits also makes it easier to troubleshoot an electrical problem; any short circuit or other problem can be isolated to the single circuit, where it can easily be diagnosed.

As noted earlier, each branch circuit should be marked showing the rooms or appliances it controls. To check all the rooms on a circuit, remove the fuse or switch the breaker for one circuit and check to see which lights or appliances are shut off; then, use a circuit tester to check switches and outlets.

Electrical Boxes and Cables

Metal electrical outlet boxes were used in older homes and are still in common use today, but some newer homes have plastic boxes.

Armored cable. Armored cable, also called BX cable, is a flexible metallic cable that contains the wiring within it, as opposed to Greenfield, which is simply the same flexible metallic conduit with the wiring installed separately. BX cable can be bent around curves or framing and is easier to install than Thinwall or other conduits. BX cable can be cut with a fine (32-point) hacksaw.

Flexible metallic BX cable or Greenfield was used in older homes to connect electrical boxes, switches, and so on. The continuous BX cable or Greenfield grounded the entire electrical system where two-slot receptacles were used, so no ground wire was necessary.

Modern wiring techniques eliminated the use of Greenfield or BX cable; plastic-sheathed electrical wiring is now used. Rather than continuous grounding through the BX cable or Greenfield, a third bare or green wire is used that is connected to either the box (if it is metal) or to grounding screws on the various devices, and then connected to the ground wire to the next wiring device. Whether you have metal or plastic boxes, the grounding wire ensures that each device is grounded.

Rigid metal conduit. Rigid metal conduit is similar to water pipe, and is used to protect wiring from weather or impact damage. Conduit is used to protect wiring both on the exterior and interior of homes. Because conduit must be threaded, special tools are needed to work it, such as pipe dies for cutting the threads, a vise for holding the conduit, and a bender to make bends in the system.

Electrical metallic tubing. Electrical metallic tubing, also called Thinwall, is lightweight conduit or piping that is used to protect exposed wiring from impact damage. Thinwall is commonly used to protect wiring that is run down a concrete basement or other masonry wall. Rather than threaded, as conduit is, Thinwall is connected via set screw–type connectors. Other types of connectors are used in commercial or other buildings, but they require special tools. You can rent or buy a bender for bending Thinwall.

Wiring

The types of wires that carry current to your home and make up the various circuits inside the home are dictated by code. The size and type of wires vary according to demanded load capacity, end use, and loca-tion. The larger the wire, the greater its load capacity is.

The size, or gauge, of wire is uniform and is set by the American Wire Gauge (AWG). Wire gauges are shown in figure 5.3. When considering the size of wire to use, keep in mind that by the AWG, the smaller the gauge, the larger the capacity. For example, most circuits in your home are either 12 gauge or 14 gauge; 12-gauge wire is larger than 14-gauge wire. Single-conductor wires are available in gauges from 8 to 20, with gauges 16, 18, and 20 used for thermostats, doorbells, and electri-cal cords.

Figure 5-3   AWG Wire Gauges

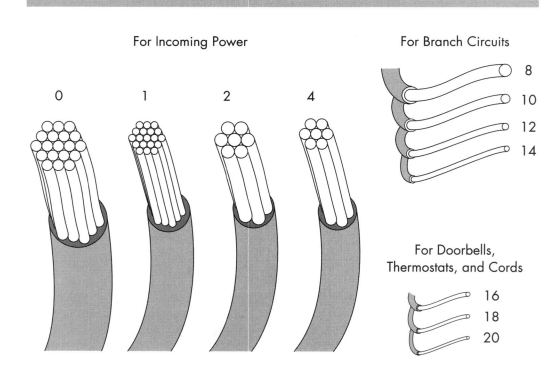

For Incoming Power

0   1   2   4

For Branch Circuits

8
10
12
14

For Doorbells, Thermostats, and Cords

16
18
20

Larger wires for incoming power (to the home) or for power to 240-volt appliances have multiple strands, and are available in gauges from 0 (the largest) down to 10, with smaller stranded wire rated from 12 down to 20. The larger stranded gauges (0 to 10) are stranded rather than solid, because they are easier to work with and bend.

In older homes, wires were encased in metallic cable. Most residential wiring used today is molded into plastic-coated cables (Romex) and contains two or more wires. This cable is called Type NMC, for non-metallic cable. The cable designation mentions the wire gauge first, then the number of wires. Common light circuits use 14-2, meaning 14 gauge, two wires. However, the cable often contains a third bare ground wire; if so, the designation is 14-2 with ground wire.

In today's homes, ordinary lighting circuits are 120 volts and use 14-2 with ground wire, protected by a 15-amp fuse or circuit breaker. Circuits in the laundry room, kitchen, workshop, or other area where electric tools or appliances are used usually require a 120-volt circuit, 12-2 with ground wire, protected by a 20-amp fuse or circuit breaker, although this depends on your local code.

Large appliances that use 240-volt power, such as water heaters, clothes dryers, or air conditioners, are wired with larger-gauge multiple-strand wires and have their own circuits. The size of the wire and amp of fuse or circuit breaker required depend on the rating of the particular appliance.

Faceplates

Faceplates are metal plates used to protect wiring from nails (e.g., wallboard or flooring nails). Faceplates have sharp points that are driven into the wood to secure them in

RULE OF THUMB

Splicing Wires

Wires may be spliced or connected in boxes only. If the wire does not reach a switch box or outlet, a junction box must be used. Codes forbid that any splice be covered up by wallboard or plaster, which would prevent inspection of the splice at a later date.

place. Where wiring runs through framing members, metal faceplates are required for Romex or plastic-coated wiring, and are optional for use with BX cable.

Switches

Switches come in a variety of types, including the single-pole switch used to control individual lights. The three-way switch is used to control a light from two separate switches located, for example, at opposite ends of a hallway. The four-way switch allows a light to be controlled by three separate switches—for example, from the bottom of a stairway, at midlanding, and at the top of the stairs.

Switches are available in various models. The common snap switch model gives a distinct snap feel or sound when the switch is turned on or off. More modern switches include the silent switch and the mercury switch. The mercury switch is a good choice; it is not only silent but much longer lasting, because it has few moving parts. Most types of switches are available with lighted handles that glow when they are turned off, so they can be easily located in the dark and can alert you that a particular light, such as an outdoor light, is off.

Other switch models include those with a switch and receptacle in one unit.

Also available are dimmer switches that can be used to replace any light switch in the home. Dimmer switches should not be used to control any appliance or receptacle, because the lower current delivered when a switch is dimmed may cause the receptacle or appliance to overheat and may cause a fire. Dimmer switches can be used to provide low-level mood lighting, or left on low to provide a night-light in a hall or bathroom. Dimmer switches also save energy, and lightbulbs will last longer when subjected to lower current. Dimmer switches designed for use with incandescent lighting should not be used for controlling fluorescent lights. Special fluorescent dimmers are available for this purpose. Dimmer switches are available for either single-pole or three-way switches. In three-way circuits you should use one ordinary three-way switch with one three-way dimmer switch.

Replacing Switches

Keep in mind that only black (hot) wires are attached to a switch, except when a white wire is marked with tape or paint to show it is used as a black wire. The white (neutral) wires are connected together with a twist-on connector and are passed through the switch box.

Replacing single-pole switches. Disconnect the power to the circuit and remove the wall plate and old switch (see figure 5.4). Reconnect the wires to the new switch. Connect the bare or green ground wire to the green screw on the switch, or to a screw in the metal box, or to the metal switch bracket. Screw the new switch to the box and replace the wall plate. Turn the power back on and test the switch.

Replacing three-way switches. Three-way switches are used in pairs to control a light from two locations. This type of switch has three wires and three wire-binding screws on the switch body (see figure 5.5). As with the single-pole switch, there will be two brass-colored terminals and a third, different-colored terminal marked "common."

To replace a three-way switch, turn off the power and remove the wall plate; then check the defective switch to see which wire is connected to the common terminal. Mark this wire with tape and remove the defective switch.

Connect the common wire to the common screw on the new switch, then connect the other two wires to the remaining silver terminals as they were connected to the old switch. If there is a green ground terminal, connect the green ground or bare wire to this terminal and replace the switch in the box. Replace the wall plate and turn on the power to that circuit. Test the switch; you should be able to turn the lights on and then walk to the other three-way switch in the pair and turn the light off from that switch.

Replacing dimmer switches. Replacing a dimmer switch, or installing one to replace either a single-pole or three-way dimmer switch, requires the same procedure as described for those switches above. (Figure 5.6 shows a single-pole dimmer switch.) Follow the directions closely. If the old switch has a ground wire, attach the ground wire to the metal box; if the box is plastic, attach the ground wire to the metal bracket on the dimmer switch.

Receptacles

Types of receptacles or outlets include two-wire receptacles (common in older homes), three-wire receptacles with a third or grounding slot, and ground-fault receptacles.

Figure 5-4 Installation of a Single-Pole Switch

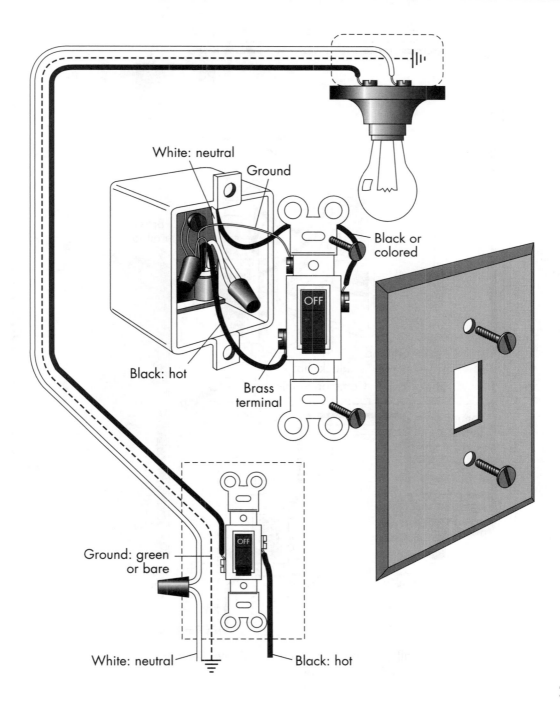

White: neutral

Ground

Black or colored

Black: hot

Brass terminal

OFF

Ground: green or bare

OFF

White: neutral

Black: hot

Figure 5-5 Installation of a Three-Way Switch

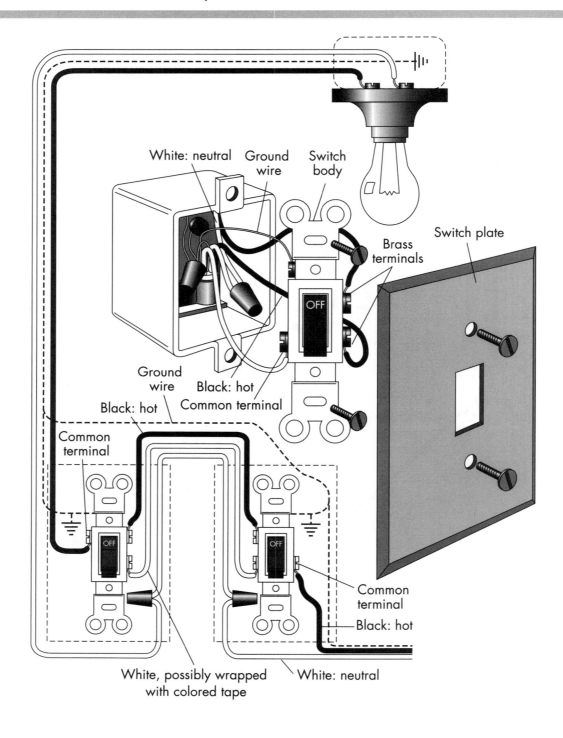

White: neutral Ground wire Switch body

Brass terminals

Switch plate

OFF

Ground wire

Black: hot

Common terminal

Black: hot

Common terminal

OFF

OFF

Common terminal

Black: hot

White, possibly wrapped with colored tape

White: neutral

Figure 5-6 Installation of a Single-Pole Dimmer Switch

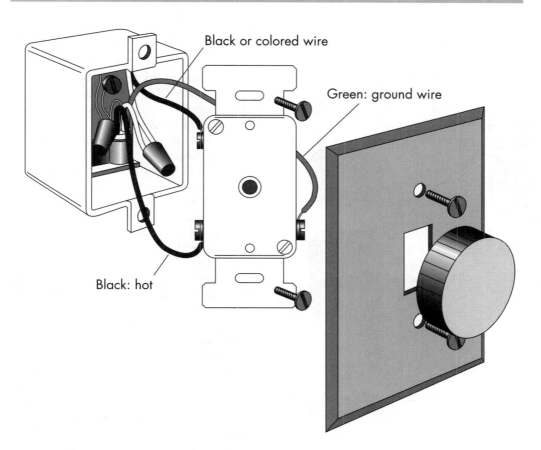

Black or colored wire

Green: ground wire

Black: hot

Replacing Receptacles

Note: Before replacing a receptacle, turn off power to the circuit and remove the wall plate. Remove the screws that connect the receptacle to the box, and check to see if there is a metal strip, called a break-off link, on the side of the receptacle. If there is, and the link has been broken, remove the same link on the new receptacle. Note also the way the wiring is connected to the receptacle. If the receptacle is the terminal, or last, receptacle on the circuit, there will be only two wires, one black (hot) and one white (neutral). (Figure 5.7

shows a terminal-circuit receptacle.) For receptacles in the middle of the circuit, there will be four wires plus a ground wire. One pair of black and white wires brings power to the outlet; the other pair of black and white wires carries the power down the line to the next receptacle. Note how many wires are attached to the receptacle, and attach the wires to the new receptacle exactly as they were attached to the old. Remember the rule that black wires are attached to brass screws, white wires to white (silver) screws, and bare or green wires to green screws, either on the receptacle or to the metal box.

Figure 5-7 Installation of a Terminal-Circuit Receptacle

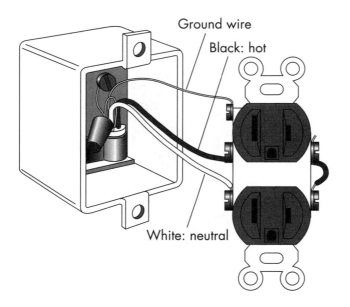

Ground wire

Black: hot

White: neutral

Most power tools manufactured within the past 10 years are double-insulated, and have plastic cases and only two-prong plugs. Some power tools are not double-insulated and have a third grounding prong, and should be used only in three-wire grounding outlets. These tools can be used in two-prong outlets if you use an adapter, provided the adapter ground wire is properly attached to a grounded box. Use a circuit tester to check if the box is grounded. Never use a two-wire adapter on any ungrounded box. Ungrounded boxes may be altered so they are grounded, but this is a job for a professional electrician. You can then install three-wire receptacles in grounded boxes, so that you can use three-prong appliance or tool plugs at that receptacle.

Ground-Fault Circuit-Interrupters

Modern electrical codes require that receptacles equipped with ground-fault circuit-interrupters be used in any area where water is present and the risk of dangerous shock is increased, for example, in kitchens, bathrooms, laundry rooms, and on outdoor outlets. In fact, homes of the future may be equipped with total ground-fault equipment, from the circuit breakers on through the home.

Ground-faults occur when tools or appliances malfunction and part of the current goes to ground. If there is no ground-fault circuit-interrupter present, the ground may travel through any person in contact with the tool or appliance. In a grounded circuit there is the same amount of current flowing through the white (neutral) wire as through the black (hot) wire. A ground-fault circuit-interrupter (GFCI) measures the current flow, and if the current in the white wire drops below the incoming current in the black wire, the device trips and shuts off the power. These units measure a drop of only 5 milliamps of current, and will trip before the person is injured.

Ground-fault circuit-interruption can be provided by GFCI-type circuit breakers, which provide ground-fault protection to all receptacles on that circuit, or by receptacle-model GFCIs. There are two types of receptacle-model GFCIs: a termination type, which protects only the outlet in which it is installed; and a feed-through model, which protects the outlet in which it is installed, plus all the down-line outlets on the same circuit.

Installing GFCI-Type Circuit Breakers

The termination-model GFCI has three wires. To install the termination model, follow the same instructions above for replacing receptacles (see page 81). The feed-through GFCI, however (see figure 5.8), has five wires, so use the following instructions.

One pair of black and white wires should be marked LINE and the other pair of black and white wires marked LOAD. If the wires are not marked, you can determine which is the line side by connecting a pair of black and white wires to the black and white wires in the box. Then tape or use a twist-on connector to join the ends of the other wires. Turn on the power and plug a lamp into the GFCI. If the lamp lights,

you have found the line-side wires. If the lamp does not light, turn off the power, disconnect that pair of wires, and connect the other pair of black and white wires to the wires in the outlet. Again, cap the loose wires and turn on the power. The lamp should light, because these wires are the line-side wires. Turn off the power and proceed.

Connect the black line-side wire to the black wire marked LINE on the receptacle. Connect the line-side white wire to the white wire marked LINE on the receptacle.

Figure 5-8 Installation of a Feed-Through Ground-Fault Circuit-Interrupter

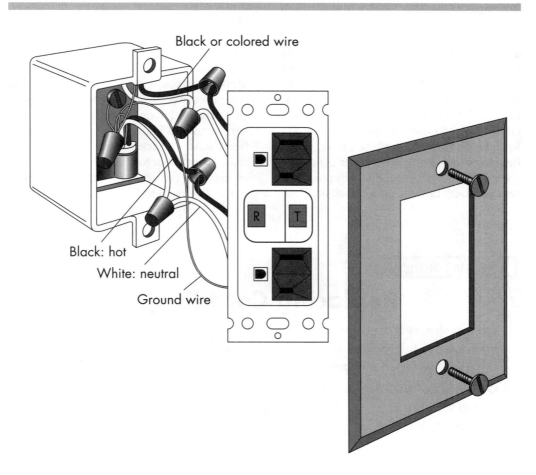

Black or colored wire

Black: hot

White: neutral

Ground wire

RULE OF THUMB

Outlet Placement

Modern electrical codes call for outlets not more than 12 feet apart, so you will never have to use an extension cord that is more than 6 feet long. The easiest method to add needed outlets is with surface wiring.

Pull together all other black wires and use a wire connector to join them to the black wire on the receptacle marked LOAD. Join all other white wires to the white wire marked LOAD on the receptacle. Connect the green receptacle wire to the green ground wire or to the grounded outlet box. (The outlet box must be grounded—use a circuit tester to test for ground. Touch one lead on the tester to the black wire, touch the other to the box. The tester light will glow if the box is grounded.) Replace the wall plate and turn the power back on. Push in the reset button of the receptacle to get power to the receptacle.

To test your installation, plug a lamp into the receptacle and push in the reset button . If the button pops out and the lamp stays on, the outlet is wired backward. Disconnect the power and recheck the wiring.

Surface Wiring

When appearance is not a factor, electrical metallic tubing, or Thinwall, can be used to provide protection to wiring. For example, Thinwall can be used to carry wiring down a concrete basement wall in a laundry room, where it is difficult to run the wire through the wall. Thinwall is often used to route wiring down garage walls, where plastic-sheathed cable wiring (Romex) might be damaged by impact from tools or other objects.

When appearance is a factor, use plastic surface wiring materials such as Wiremold. Wiremold is often used to add an extra outlet on a wall, or to extend wiring from a wall outlet up to a ceiling fan. These surface wiring systems offer shallow outlet boxes that extend only $1\frac{1}{4}$ inches from the wall or ceiling. Flat plastic channels are cut to fit between the existing

RULE OF THUMB

Replacing Cord Plugs

On a polarized cord (one that has a neutral blade wider than the black, or hot, side), be sure to attach the black wire to the brass screw. On flat cords, one side of the wire will have a ribbed surface; the ribbed wire is the white (neutral) wire.

On a three-wire cord, connect the black wire to the brass screw, the white wire to the silver screw, and the green wire to the ground screw. This will ensure that polarity is preserved in the cord.

or source outlet and the new outlet location. The wires are run through the channels, then snap-on plastic covers are used to provide a smooth and pleasing appearance. Special molded-plastic angle or corner pieces can be used to turn corners or change channel direction. The plastic channels can be painted or stained to blend them into the surface on which they are mounted.

Appliance and Extension Cords

For safe operation, be sure to inspect appliance and extension cords frequently for wear and damage.

When replacing cord plugs, do not depend on the wire screws alone to secure the wire; provide extra strain relief capacity.

Foolproof Advice
for Electrical Cord Safety

- Repair or replace any damaged cords.
- Look for nicks or cracks in the cord insulation, and check to be sure plugs are well secured and the plug blades are not bent.
- If the plug blades are loose in the receptacle, do not bend the blades, but first check the plug blades in another receptacle to be sure the problem is in the plug blades and not in the receptacle spring contacts.
- Bend or adjust the plug blades only when they are obviously bent.
- Replace badly damaged plugs.
- Check three-prong grounded extension cords with a continuity tester to be sure the wires are intact.
- Always pull the plug, not the cord, when disconnecting a cord.
- Disconnect a cord from the outlet first, then from the appliance. Don't run extension cords under rugs or other materials, where they may become heated and cause a fire.
- Don't run cords across doorway thresholds, where they are subject to wear and abuse.
- Don't use lightweight cords to run power tools; use an extension cord that has enough capacity to handle the load. Note that there will be a line drop (drop in delivered voltage) when using long extension cords, because long cords provide a higher resistance to current flow.

Strain relief can be provided by using plugs that clamp a portion of the wire beyond the wire screws. If the plug does not have a clamp, then tie a figure-eight knot in the end of the wire to keep it from pulling out of the plug. This type of knot is called an underwriter's knot because the Underwriters' Laboratory (UL) developed the technique.

Low-Voltage Circuits: Doorbells

Some systems in your home, such as doorbells and thermostats, require lower voltage than your 120-volt circuits provide, so these devices have transformers to reduce, or step down, the line voltage to 10 to 24 volts. Voltage ratings of doorbells may vary, so be sure the transformer you use matches the rating of the doorbell.

Repairing Doorbells

To troubleshoot doorbell problems, first use a circuit tester to test whether power is reaching the transformer. If there is power to the transformer, remove the doorbell button and place a screwdriver blade across the contacts. If the bell rings, put in a new button. If the bell does not ring, check the bell or chime by running a wire between the transformer terminal and the button terminal. If you get a spark but the bell doesn't ring, replace the bell. If there is no spark, check the wiring from the transformer to the bell. If the bell works but shorting the wire at the button doesn't cause the bell to ring, check the wiring from the button to the bell.

Outdoor Wiring

Outdoor wiring is exposed to severe conditions, such as moisture and extremes of heat and cold. For proper operation and to ensure safety, you must observe special techniques and use weatherproof devices and wiring when working on outdoor wiring projects.

The *National Electrical Code* requires that all outdoor wiring be protected by ground-fault circuit-interrupters, either at the circuit breaker or at receptacle-model GFCI devices. At outlets, use three-wire grounded receptacles and a weatherproof outlet cover.

To add outdoor outlets. Run the power either from an indoor box on an outside wall or from a junction box in the house, garage, or basement. Choose a source box that is nearest the location of the new outlet. Remember always to shut off power to the source outlet, and follow the usual wiring rule of attaching black wires to brass screws, white wires to silver screws, and green or bare wires to green grounding screws.

To install porch or eave lighting. Access power from the nearest indoor outlet and use outdoor fixtures and bulbs. Turn off the power to the source box and install the outdoor box and switch box. Run plastic cable (Romex) wire to the switch box and to the outdoor box. Make the connections and turn the power on to test the switch and light.

To install a lamppost. See figure 5.9 for an outdoor lighting installation. Select an existing circuit, or source box, inside the home that is nearest the new lamppost. This will probably be an outlet or junction box in the basement. Turn off the power to the source box. Install an indoor switch at the desired location; then, from the source box to the outside, run either PVC plastic conduit (similar to plastic water pipe) or metal conduit if your local building codes require metal. Dig a 12-inch-deep trench from the metal conduit to the lamppost. Attach a 90-degree elbow on the outside end of the conduit, and install a drop conduit so it extends beneath ground level in the trench.

Figure 5-9 Outdoor Lighting Installation

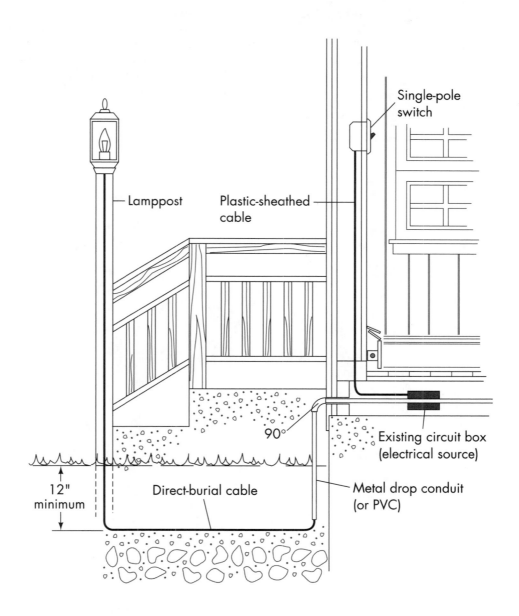

Single-pole switch

Lamppost

Plastic-sheathed cable

90°

Existing circuit box (electrical source)

12" minimum

Direct-burial cable

Metal drop conduit (or PVC)

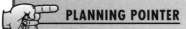

PLANNING POINTER

Easy Outdoor Lighting

The easiest way to install outdoor lighting is to use low-voltage outdoor lighting kits. Various lighting kits are available that can be safely plugged into any outdoor outlet.

For the wiring, use UF cable, a plastic-sheathed cable that can be buried underground. Run the cable from the source box, through the conduit, across the trench, and up to the post top. Run the wire to the switch box. Attach the wires to the source wires and to the post light, attaching the black wire to the brass screw, the white wire to the silver screw, and the green or bare ground wire to the ground screw. Turn on the power and test the light.

6 Painting

Paint is the finish that protects and decorates most of the surfaces inside and outside a house. The surfaces range from protected interior walls to unprotected exterior siding, which has to withstand extremes in climate and conditions. Here are the basics for choosing the right type of paint for the protection and finish wanted.

Proper preparation of the surface is the key to a lasting paint job. To help you get the most out of any paint, the proper application of primer-sealers, stain killers, and other coatings is discussed.

Good advice on the application of exterior stains and a complete trouble-shooting section will also arm you with the knowledge to tackle any painting or staining project around your house.

Paints and Primer-Sealers

Types of Interior Paint

There's a right kind of paint for every application to produce the best-looking results and protection.

Latex paint. Latex paint is water soluble, so it's easy to apply and dries quickly, often an important feature if more than one coat is required in a day.

Cleanup involves washing with soap and water—a relatively simple chore. Paint spills can be wiped away with a damp rag, so it's a good choice for a first-time painter. If you're painting a metal surface with latex, you must first use a rust-preventive primer because latex, by its nature, can cause rust on an

RULE OF THUMB

Painting with Latex

When using latex paints, clean your brush at least every 4 hours, because latex dries fast and will harden in the heel of the brush as you use it. Once hardened, it's more difficult to clean at the end of the day.

unprotected surface. Latex paint comes in a range of finishes (lusters), from flat to semi-gloss.

Flat latex is primarily used on interior walls and ceilings, while shinier finishes such as eggshell, satin, and semigloss are used on woodwork and trim because they dry to a hard finish.

Alkyd paints. Alkyd paints are solvent-based and thus require more diligence to apply. Spills and cleanup involve using paint thinner. But the paint is a proven choice of many professionals because it provides a hard, durable finish and flows on evenly. It's a popular choice for woodwork and trim requiring frequent washings. Like latex, alkyd paint comes in a full range of finishes, from gloss to flat.

Enamel paint. Enamel paints are high-gloss and are available with either a latex or alkyd base. Alkyd-based enamels tend to dry slower and harder than latex products and are the best choice for interior woodwork. Latex-based enamels are a good choice for exterior work. These tend to hold their gloss longer and chalk less than alkyd paints. The most important factor in the performance of either type enamel is its initial quality. A top-of-the-line alkyd paint will perform better than a bargain-basement latex product and, of course, the reverse is also true. A top-

PLANNING POINTER

Choosing Paint

A good combination is using flat latex paint on walls, and alkyd paint on woodwork because it dries to a hard finish.

quality latex enamel will outlast a bargain alkyd-based product.

Other types of interior paint. Special-use alkyd and latex paints are designed for floors, textured and flat paints for ceilings, and mildew-proof and waterproofing paints for moist areas, basement walls, and metal surfaces.

Primer-Sealers and Stain Killers

A primer-sealer is a finish that prepares the surface for paint, either latex or alkyd, with an undercoating. It seals the surface to ensure a more uniform appearance so the paint doesn't soak into it. It's required on porous surfaces such as unfinished wood or new wallboard, on slick surfaces like tile or paneling and gloss-painted woodwork, and on stained and damaged surfaces like knotty wood or ink marks on a wall. Primer-sealer, which is usually white, can be applied as is or tinted by the retailer to match the top coat of paint. Table 6.1 shows which primer and top coat to use on various surfaces.

Primer-sealers are oil-based, water-based, and shellac-based, which is the general-purpose primer for all types of surfaces and stains. The shellac- and water-based products can be used as a stain killer. Ballpoint pen ink, crayon marks, and most other wall stains will bleed through latex or

RULE OF THUMB

Proper Ventilation with Alkyd Paint

When applying alkyd paint, allow for plenty of ventilation. If you're sensitive to the synthetic resins used in the paint, wear a respirator to protect you from the fumes.

Table 6.1 Primer-Top Coat Use

Surface	Primer	Top Coat
Unpainted plaster	Latex/alkyd	Flat latex
Unpainted wallboard	Latex/alkyd	Flat latex
Painted plaster	None	Flat latex
Painted wallboard	None	Flat latex
Unpainted wood trim	Alkyd	Alkyd–gloss/semigloss enamel
Painted wood trim	None	Alkyd–gloss/semigloss enamel
Unpainted wood floor	Alkyd	Alkyd–gloss/semigloss enamel
Painted wood floor	None	Alkyd–gloss/semigloss enamel
Acoustical tile	Latex	Flat latex
Brick	Latex	Alkyd/latex–gloss/semigloss enamel
Masonry	Latex	Alkyd/latex–gloss/semigloss enamel
Steel	Rust-inhibiting primer	Alkyd–gloss/semigloss enamel
Aluminum	None	Alkyd–gloss/semigloss enamel
Galvanized metal	Zinc dust primer	Alkyd/latex–gloss/semigloss enamel
Cast iron	Red Lead/Zinc Cromat	Alkyd/latex–gloss/semigloss enamel

oil-based paints unless the stain is first coated with a stain killer.

Use a shellac-based primer-sealer for sealing knots and sap streaks in new wood, such as the ones found on new molding and doors. It seals out grease splatters, water stains, and smoke damage as well as odors. Because it contains alcohol solvent it dries quickly and brushes clean with household ammonia and water. Table 6.2 shows the primer to use on various hard-to-paint surfaces.

What About Lead Paint?

The Consumer Product Safety Commission says that lead-based paint is hazardous to your health. It's a major source of lead poisoning for children and can affect adults. About two-thirds of the homes built before 1940 and one-half of the homes built from 1940 to 1960 contain heavily leaded paint. It may be on any interior or exterior surface, particularly woodwork, doors, and windows.

Table 6.2   Primers to Use on Hard-to-Paint Surfaces

Surface	Primer
Smoke, fire damage	Shellac stain killer/oil-based primer
Water stains	Shellac stain killer/oil-based primer
Water stains–acoustical tile	Shellac stain killer/oil-based primer
Wood knots, sap	Shellac stain killer
Lipstick, grease, crayon	Shellac stain killer/oil-based primer
Ballpoint, marking pen	Shellac stain killer
Exterior tar, asphalt	Water-based acrylic primer
Creosote	Shellac stain killer
Graffiti	Shellac stain killer/oil-based primer
Nail heads, rust	Water-based stain killer/oil-based primer
Rust stains	Water-based stain killer/oil-based primer
Mildew	Shellac stain killer/oil-based primer

The danger of lead paint is its dust particles, which infiltrate a house. The simple act of window sashes opening and closing creates chips and scrapings that can be harmful. When years of paint are scraped and sanded, the dust settles on the ground surrounding the house, then blows inside through open doors and windows.

Testing for Lead

Testing for lead in a home's paint is particularly important if infants, children, or pregnant women are present. To test for lead, there are various do-it-yourself lead-testing kits sold at home centers, hardware stores, and paint retailers, and there are professional lead-testing services. In some states the health department offers the testing service to its residents.

Contact your local state health department for its policy about lead abatement. If you have a major lead hazard, hire a qualified contractor. Your strategies for abatement include:

- Encapsulating lead-painted walls with a permanent barrier, such as a wall liner, wallpaper, paneling, or an encapsulant coating; on a lead-painted floor, install new vinyl sheet flooring or tile or wood flooring.

- Replacing the lead-painted surfaces with new ones; this can mean replacing wallboard, moldings, doors and windows, and exterior siding.

Removing Lead Paint

When removing lead paint with chemicals, the stripping creates a dangerous situation because of the caustic fumes. Consider removing things like the molding and doors to have them commercially stripped. Another option is to hire a professional to do the job.

You can rent a HEPA (high-efficiency particle air) sander to catch lead dust from flat surfaces like walls and floors, and an electric heat gun to remove heavy layers of paint, but this creates and spreads lead fumes with the dust. When using any of this equipment, you should wear an approved respirator, heavy neoprene gloves, goggles, and a long-sleeved shirt and pants to cover your skin.

When any lead abatement work is done, all children and household members should leave the premises until the work is completed and the work area is cleaned.

To dispose of the debris, contact your local health department for directions.

Foolproof Advice for Environmentally Friendly Disposal of Paint

The best way to dispose of any paint is to use it up. Contact your local sanitation department to see if your community has an active paint reuse/recycling program. Don't throw away leftover paint in your trash.

- Wash latex paintbrushes and tools in your sink, but not near a storm sewer drain.

- Don't empty alkyd paint and thinner into storm sewers, household drains (especially if you have a septic tank), or on the ground.

- Pour unused latex paint into an absorbent material, such as kitty litter, shredded newspaper, or sawdust. Let it dry completely, then dispose of the dried material in your regular trash.

- Save alkyd paint for a hazardous-waste collection program, or ask your local sanitation department how to dispose of it.

- Reuse paint thinner by letting it sit in a closed container until the paint particles settle to the bottom. Pour off the clear liquid into a clean container with a lid for reuse. Add an absorbent material like kitty litter to the residue; then, when it's dry, dispose of it in your trash.

- To clean paint rollers and brushes, buy a paint spinner that spins them clean after each use. The tool works like a kid's toy top to spin out excess water or paint thinner used to clean rollers and brushes.

Tools and Equipment for Interior Painting

Brushes and Rollers

Buy the best-quality brushes and rollers you can afford. Table 6.3 outlines brush use for nylon/polyester and natural-bristle brushes. Here's how to choose the best ones for the job.

Nylon or polyester brushes. These brushes are suitable for latex or oil paints, so if you're buying only one brush, make it one with nylon or polyester bristles.

Natural-bristle brushes. Natural-bristle brushes are more expensive, and work best with alkyd paint. They hold the paint, so you have better control over its flow. There's less dripping, and the brush applies the paint so it dries to a smooth finish. Don't use a natural-bristle brush with latex paint; it goes limp.

Foam brushes. These are designed for short or quick touch-up jobs. Used with latex paint, foam brushes can be washed out and reused, but they're designed as throwaways.

Sash brushes. The bristles of sash brushes are cut at an angle for painting thin areas like window sashes.

Typical square-cut or straight-across brushes. These are useful for cutting in, or for painting the area next to windows and doors or in the corners and at the ceiling of a room.

Brushes with elongated pencil-style handles. These brushes give the painter a good grip for doing detail work.

Paint rollers and sleeves or covers. These are sized 7 or 9 inches wide. A 9-inch roller with an extension pole that screws in the end of the roller handle is a good rig for painting. Choose a roller with a handle frame stiff enough to resist flexing and with a sleeve that can be slid on and off with ease. Table 6.4 shows roller cover use. Nap height refers to the length of the roller nap or how fuzzy the roller is. The higher the nap, the faster the roller will paint but the rougher the finish.

Table 6.3 Brush Use

Nylon/Polyester	Natural-Bristle	Size	Uses
Latex, water-based paint, enamels	Oil-based paint, varnishes, lacquers	1"	Narrow trim, shutters, touch-up
		1½" to 2"	Window sashes, narrow door casings, thin trim, shutters
		2½" to 3"	Cabinets, baseboards, shelves, doors, fences, gutters, stairs, steps
		3½" to 4"	Walls, ceilings, siding, paneling, exterior fascia and trim

Table 6.4 Roller Cover Nap Height

Nap Height	Uses	Paint
⅜" to ¾"	Smooth walls/shelving/cabinet work	Semigloss enamels
½"	Lightly textured/previously painted walls	Eggshell/flat
¾"	Rough/patched walls/walls with many coats of paint	Flat
1" or more	Textured surfaces	Flat

Paint pans. Paint pans come in a variety of materials. Choose a sturdy metal or heavy plastic paint pan with legs or corner brackets that hook onto the top of a ladder.

Paint pads. Paint pads are plastic holders with a removable synthetic pad and adjustable trim guides for applying paint without a brush and roller. They vary in size, usually 4 to 10 inches wide, with a straight edge at the top. There are also special-use pads designed for painting corners and under the lip of house siding.

Spray Equipment

For many painting chores, such as painting large, uneven, rough, or textured surfaces, the best choice for paint application is spraying. Most home-painting chores can be done either with an aerosol can or with a handheld paint sprayer. For larger painting jobs, you can rent air-driven or airless spray equipment of varying capacity. Some units can hold only 1 gallon of paint, and some units draw paint from a 5-gallon pail. Be sure to tell the rental operator the kind of material you will be spraying so that you'll be provided with the proper nozzle orifices to handle your material.

RULE OF THUMB

Painting Equipment

To save money, buy good painting equipment and take care of it. Start by buying paint rollers with a waterproof center core. Plastic-coated cardboard or solid plastic cores will stand up to many cleaning cycles, compared with the less expensive, throwaway, plain cardboard ones that will come apart after a few uses. Choose a few polyester-bristle brushes and clean them thoroughly after every use.

Air-Driven Equipment

Air-driven spraying equipment consists of an air compressor and air and material hoses to the paint gun, a paint pot to hold the paint, and the gun itself. With this system, the amount of air pressure and the type of nozzle orifice are selected to match the requirements of the material being sprayed. Air pressure to the paint pot pushes the paint up the material hose to the gun. At the gun nozzle, the air atomizes the paint or other material and the gun nozzle is then adjusted to provide the right spray pattern. Air-driven paint equipment often is the pro's choice for spraying most alkyd-based paints and metal paints such as automotive finishes, because airless paint sprayers leave a slight, orange-peel effect on the paint.

Compressor service. Because air is pulled into the compressor and is mixed with paint at the gun nozzle, any moisture in the air can collect in the compressor tank. The water can mix with alkyd paints and ruin the paint finish. Remove the drain plug on the bottom of the air compressor and drain out any water after each use.

Airless Spray Equipment

Airless spray equipment consists of an airless unit and paint pump with a hose that connects to the gun. Airless sprayers actually are electric high-pressure pumps that pump the paint through a single hose to the gun. At the gun the pressurized paint is forced through a tiny orifice in the nozzle, and the pressure on the paint forces it to atomize into a fine spray mist. Airless paint sprayers are also available as a handheld sprayer that incorporates a small material pump with a paint reservoir attached. Airless sprayers are a popular choice for painting with thicker-bodied latex paints, but by choosing the right orifice or tip, they can also be used for applying exterior oil-base stains.

Adjusting the Spray Pattern

As mentioned, the spray pattern depends on the amount of air pressure, the viscosity or thickness of the material being sprayed, and the type of nozzle orifice through which the paint is driven. Adjust the air pressure per the manufacturer's directions.

Follow the paint manufacturer's directions for thinning the paint for spraying. Be

sure you have the right nozzle on the gun. When you have the right paint viscosity, air pressure, and nozzle on the gun, test-spray to check the spray pattern. Spray the material on a large piece of cardboard—a large appliance-shipping box makes an excellent spray paint booth. Before beginning to spray your project, adjust the gun nozzle until you get the proper spray pattern.

Spraying Paint

To spray paint, hold the gun nozzle at a distance from the surface to be painted so that it produces the desired pattern. Holding the gun constantly at that distance from the work piece, move the gun in an arc across the surface. If you stop the movement of the gun, you must release the trigger, because holding the gun in one spot with the trigger held back will load up that spot with excess paint and produce paint runs. Do not worry about covering the surface at one pass; keep the gun moving and let the paint set slightly before spraying over the same area again. Repeated light coats will eventually build the paint film and will result in complete coverage of the project.

Cleaning Spray Equipment

Failure to properly clean spray equipment can either clog or ruin the equipment when paint or other materials harden inside the pumps, hoses, or gun. Follow the manufacturer's directions for cleaning any spray equipment. In general, after finishing spraying, empty the paint pot and fill it with the appropriate solvent: water for latex or water-based materials, turpentine or mineral spirits for alkyd-based paints. Then spray the solvent through the sprayer until you see no paint residue at the nozzle.

Spraying solvent through the sprayer may be sufficient to clean some equipment. For other equipment you must remove the

Safety Tips for Using Spray Equipment

- Because paint is atomized under pressure, the tiny paint particles drift through the air and can be inhaled. Use an approved spray respirator and eye goggles when spraying paint.

- Cover your body completely with a long-sleeved shirt and painter's pants and hat so skin is minimally exposed to the paint.

- Do not eat, drink, or smoke while spraying paint.

- Wash your hands with a good cleaner such as GoJo after painting and before touching food or beverages.

- **Never point a spray gun at a person.** The pressurized paint from an airless sprayer can actually penetrate the skin.

nozzle or tip from the spray gun and clean these parts separately in the appropriate solvent.

Drop Cloths

Protect the room and everything in it with drop cloths. Professionals use heavy canvas drop cloths, which are sold in room sizes ranging upward of 9 by 12 feet and as runners for hallways that are 6 feet wide and of various lengths. Canvas drop cloths lie flat while covering the floor and withstand heavy foot traffic and paint spills.

Heavy plastic drop cloths are a less expensive choice, but they're not as durable as canvas. Paper-faced plastic drop cloths are the best choice for the casual painter. This type of drop cloth is inexpensive, lies flat, and is safe to walk on. Given reason-able care, paper-faced drop cloths can be reused many times.

If you drip paint on a plastic drop cloth, wipe up the paint because globs of wet paint can be tracked throughout the house, damaging the flooring.

PLANNING POINTER

Choosing a Drop Cloth

To size a room for a drop cloth, add at least 2 feet to the length and width of the room. You'll need excess drop cloth to gather at the sides of the room so the corners and sides are fully protected.

RULE OF THUMB

Easy Room Painting

For painting more than one room in the same color, use a 5-gallon paint bucket with a lid and a roller screen that hangs in the bucket. The screen hooks over the bucket's edge and hangs into the paint. Fill the bucket with paint and work the roller up and down the screen with the paint. Get an extension pole for the roller handle so you don't need a ladder to roll the ceiling.

To paint a room with a window and door, the tools you'll need are 1-inch and 2½-inch sash brushes for the trim, a 2½-inch straight brush for cutting in paint in the corners and at the ceiling, and a roller, sleeve, and pan for the walls and ceiling.

Thin plastic drop cloths should be used for draping to protect furnishings that can't be moved from paint splatters. They can be reused several times, but check them carefully for small holes before reusing. Don't use these drop cloths on the floor, where they are slippery to walk on.

RULE OF THUMB

Avoid Painting Yourself into a Corner

If you're painting a floor, paint yourself out of the room by beginning in a far corner and working toward the door. If you're painting stairs, paint every other riser and tread so you'll have use of the stairs while the paint dries.

Masking Tape

To protect woodwork and trim, fireplaces, and anything else from paint, use masking tape, which is a craft paper tape that comes in various sizes beginning at 2 inches wide. One edge has adhesive to apply to the protected surface. Cut and shape the tape to cover the surface, and then when the paint has dried, give the tape a gentle tug to remove it.

Interior Painting Techniques

Painting Walls and Ceilings

Paint the ceiling by cutting in, or outlining, the ceiling with paint with a wide brush, creating a 2-inch-wide band of wet paint.

PLANNING POINTER

Room-Painting Sequence

To get the best-looking results and create a work pattern that flows easily, there's a sequence to the job of painting a room. The order of painting tasks is important so there's no repetition or corrective measures needed. Sequence is particularly important with wet paint, because no one wants to drip ceiling paint on freshly painted walls or paint a window closed. Therefore, paint the ceiling first, then the walls, then the doors, windows, and woodwork.

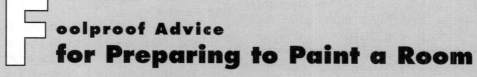

oolproof Advice
for Preparing to Paint a Room

- Remove all the wall decorations and as much furniture from the room as possible.
- Stack the pieces that remain in clusters away from the walls. Allow as much space as possible away from the walls so there's room to comfortably use a paint roller.
- Protect the floor and furnishings with drop cloths.
- Remove the plate covers from electrical receptacles and switches.
- Cover large chandeliers or light fixtures by wrapping them in large plastic trash bags and securing with masking tape.
- Dust the area to remove cobwebs, and wash away any grease or surface dirt.
- If there's dried paint from a previous job, remove it with a green scouring pad designed to scrub pots and pans. First try using water to see if it was latex paint; if it doesn't budge, use mineral spirits. Do this very gently so you don't damage the surface.
- Fill all nail holes and minor cracks with a joint compound and sand it smooth.
- When painting, wear comfortable old clothes and shoes that won't be ruined with a smudge of paint. Wear a hat to protect your hair, especially if you're painting a ceiling.

Then use the roller to paint the surface of the ceiling, feathering the paint in with the outline of paint.

Paint the walls by first outlining the walls at the ceiling with wall paint and then paint the corners and around the window and door trim. Then use a roller to paint the surface of the walls.

When the wall paint has dried, paint the woodwork, trim, windows, and doors.

 **RULE OF THUMB**

Estimating Time

It takes the average person 1 hour to paint 1,000 square feet, plus 1 hour for each window or door. The higher the roller nap, the faster you can paint. A high-napped roller holds more paint and you don't have to reload it as often as a low-napped one. It does, however, produce a stippled finish.

Painting Windows

Use a 1½-inch brush to paint from the inside out beginning at the muntins, the thin vertical and horizontal dividers between the panes of glass in double-hung windows. (If you have casement or double-hung windows with removable muntins, take them out and begin on the frame nearest the glass.)

If the upper sash of a double-hung window is movable, reverse the position of the inner (or lower) sash and the outer (or upper) sash. First paint the lower half of the outer sash and then paint the inner sash. Don't paint the top edge where the lock is until the last.

Don't close the window completely but move the sashes back to their regular locations. Then paint the tops of the outer and inner sashes.

Use a wider, 2½-inch brush to paint the frame again, working from the inside out and moving the sash cord out of the way. Paint the window casing, then the sill and the apron, which is the trim beneath the windowsill.

Because of its hard finish and washability, alkyd paint is a good choice.

RULE OF THUMB

Painted Window Hardware

If a window lock is covered with heavy layers of paint but works properly, leave it alone. If you remove the lock, you'll crack the existing paint and leave a dent in the paint in the shape of the lock. Unless you replace the lock with the same type, you'll have to sand away the buildup of paint.

Foolproof Advice
for Preparing to Paint Woodwork

- Degloss shiny surfaces with a chemical solvent deglosser. It removes dirt and grease as well as the sheen.

- Use 120-grit sandpaper on flat surfaces with a sanding block or electric palm sander to smooth out chipped areas. Make sure to feather the rough areas smooth.

- Fill small holes and cracks with a wood filler or joint compound, making sure to sand it smooth and refill any indentations.

- Protect window glass from paint with masking tape.

- Loosen door locks from interior doors or mask around them with masking tape.

- Remove locks and window hardware that will be replaced and fill in the holes with wood filler. (If hardware is not painted, it may be easier to remove it rather than paint around it.)

Painting Doors

If the door is flat, use a roller to paint the surface first, then use a brush to paint the door's sides, top, and bottom edges. Avoid dripping paint, and be careful cutting in around the door hinges. Paint the doorjamb last.

Paint a panel door like a window, from the inside out. Use a 2½-inch brush and paint a small section at a time, working on individual panels. Paint the decorative edge by outlining paint around the molding of each panel. Work from the top down and paint the face of the panel. Work the paint into the outline area to prevent creating lap marks. Paint the interior of all the panels, then paint the stiles, or vertical partitions, between the panels. For a smooth finish, carefully feather out the paint at the intersection where the stiles and rails, or horizontal partitions, meet. Finally, paint the top and bottom rails, then the outside stiles to complete the surface of the door.

Then paint the door's side, top, and bottom edges, being careful to cut in around the hinges and lock mechanism. Paint the doorjamb last.

Because of its hard finish and washability, alkyd paint is a good choice.

Estimating How Much Paint You'll Need

The average room does not require a lot of trim paint. A quart will usually be enough. If you figure you'll need 3 quarts, buy a gallon.

Most trim paints cover between 350 and 400 square feet of surface per gallon. Painters allow about 8 square feet of paintable area for each window and about 25 square feet of area for each door. Extra paint is needed for the window's trim, or casing, and its jamb.

Tables 6.5 and 6.6 show the amount of paint you will need to paint the walls and

Table 6.5 Gallons of Paint to Purchase—Ceilings[a]

Room Length											
20	0.4	0.5	0.5	0.6	0.6	0.7	0.7	0.8	0.8	0.8	0.9
19	0.4	0.5	0.5	0.5	0.6	0.6	0.7	0.7	0.8	0.8	0.8
18	0.4	0.4	0.5	0.5	0.6	0.6	0.6	0.7	0.7	0.8	0.8
17	0.4	0.4	0.5	0.5	0.5	0.6	0.6	0.6	0.7	0.7	0.8
16	0.4	0.4	0.4	0.5	0.5	0.5	0.6	0.6	0.6	0.7	0.7
15	0.3	0.4	0.4	0.4	0.5	0.5	0.5	0.6	0.6	0.6	0.7
14	0.3	0.3	0.4	0.4	0.4	0.5	0.5	0.5	0.6	0.6	0.6
13	0.3	0.3	0.3	0.4	0.4	0.4	0.5	0.5	0.5	0.5	0.6
12	0.3	0.3	0.3	0.3	0.4	0.4	0.4	0.5	0.5	0.5	0.5
11	0.2	0.3	0.3	0.3	0.3	0.4	0.4	0.4	0.4	0.5	0.5
10	0.2	0.2	0.3	0.3	0.3	0.3	0.4	0.4	0.4	0.4	0.4
9	0.2	0.2	0.2	0.3	0.3	0.3	0.3	0.3	0.4	0.4	0.4
Width	10	11	12	13	14	15	16	17	18	19	20

[a] Table assumes coverage of 450 square feet per gallon.

Table 6.6 Gallons of Paint to Purchase—Walls[a]

Room Length	Height																	
	7.5	8	9	7.5	8	9	7.5	8	9	7.5	8	9	7.5	8	9	7.5	8	9
20	0.9	1.0	1.1	1.0	1.0	1.2	1.0	1.1	1.2	1.0	1.1	1.2	1.1	1.1	1.3	1.1	1.2	1.3
19	0.9	1.0	1.1	0.9	1.0	1.1	1.0	1.0	1.2	1.0	1.1	1.2	1.0	1.1	1.2	1.1	1.1	1.3
18	0.9	0.9	1.0	0.9	1.0	1.1	0.9	1.0	1.1	1.0	1.0	1.2	1.0	1.1	1.2	1.0	1.1	1.2
17	0.8	0.9	1.0	0.9	0.9	1.0	0.9	1.0	1.1	0.9	1.0	1.1	1.0	1.0	1.2	1.0	1.1	1.2
16	0.8	0.8	1.0	0.8	0.9	1.0	0.9	0.9	1.0	0.9	1.0	1.1	0.9	1.0	1.1	1.0	1.0	1.2
15	0.8	0.8	0.9	0.8	0.8	1.0	0.8	0.9	1.0	0.9	0.9	1.0	0.9	1.0	1.1	0.9	1.0	1.1
14	0.7	0.8	0.9	0.8	0.8	0.9	0.8	0.8	1.0	0.8	0.9	1.0	0.9	0.9	1.0	0.9	1.0	1.1
13	0.7	0.7	0.8	0.7	0.8	0.9	0.8	0.8	0.9	0.8	0.8	1.0	0.8	0.9	1.0	0.9	0.9	1.0
12	0.7	0.7	0.8	0.7	0.7	0.8	0.7	0.8	0.9	0.8	0.8	0.9	0.8	0.8	1.0	0.8	0.9	1.0
11	0.6	0.7	0.8	0.7	0.7	0.8	0.7	0.7	0.8	0.7	0.8	0.9	0.8	0.8	0.9	0.8	0.8	1.0
10	0.6	0.6	0.7	0.6	0.7	0.8	0.7	0.7	0.8	0.7	0.7	0.8	0.7	0.8	0.9	0.8	0.8	0.9
9	0.6	0.6	0.7	0.6	0.6	0.7	0.6	0.7	0.8	0.7	0.7	0.8	0.7	0.7	0.8	0.7	0.8	0.9
Width	10			11			12			13			14			15		

Room Length	Height														
	7.5	8	9	7.5	8	9	7.5	8	9	7.5	8	9	7.5	8	9
20	1.1	1.2	1.4	1.2	1.2	1.4	1.2	1.3	1.4	1.2	1.3	1.5	1.3	1.3	1.5
19	1.1	1.2	1.3	1.1	1.2	1.4	1.2	1.2	1.4	1.2	1.3	1.4	1.2	1.3	1.5
18	1.1	1.1	1.3	1.1	1.2	1.3	1.1	1.2	1.4	1.2	1.2	1.4	1.2	1.3	1.4
17	1.0	1.1	1.2	1.1	1.1	1.3	1.1	1.2	1.3	1.1	1.2	1.4	1.2	1.2	1.4
16	1.0	1.1	1.2	1.0	1.1	1.2	1.1	1.1	1.3	1.1	1.2	1.3	1.1	1.2	1.4
15	1.0	1.0	1.2	1.0	1.1	1.2	1.0	1.1	1.2	1.1	1.1	1.3	1.1	1.2	1.3
14	0.9	1.0	1.1	1.0	1.0	1.2	1.0	1.1	1.2	1.0	1.1	1.2	1.1	1.1	1.3
13	0.9	1.0	1.1	0.9	1.0	1.1	1.0	1.0	1.2	1.0	1.1	1.2	1.0	1.1	1.2
12	0.9	0.9	1.0	0.9	1.0	1.1	0.9	1.0	1.1	1.0	1.0	1.2	1.0	1.1	1.2
11	0.8	0.9	1.0	0.9	0.9	1.0	0.9	1.0	1.1	0.9	1.0	1.1	1.0	1.0	1.2
10	0.8	0.8	1.0	0.8	0.9	1.0	0.9	0.9	1.0	0.9	1.0	1.1	0.9	1.0	1.1
9	0.8	0.8	0.9	0.8	0.8	1.0	0.8	0.9	1.0	0.9	0.9	1.0	0.9	1.0	1.1
Width	16			17			18			19			20		

[a] Table assumes coverage of 450 square feet per gallon.

RULE OF THUMB

Leftover Paint

After painting a room, save the leftover paint in case it's needed to touch up an area that's been damaged and repaired. If a small amount of paint is left over, store it in a container with a tight-fitting lid. Store larger amounts of leftover paint in the paint can, after cleaning the lip of the can so it's free of dried paint. Use a hammer to seal the lid. Label the paint, noting the room where it was applied, like LIVING ROOM WOODWORK, and the date it was used.

ceiling of a given size of room. To use these tables, read up the left side of the table to find the number closest to the length of the room and then along the bottom of the table to find the width. Read the number of gallons at the intersection. Table 6.6 has three columns for each wall height. Choose the height closest to your actual room. Convert the decimal part of the gallon figure to gallons or quarts. Round any decimal over .5 to the nearest gallon. In most cases, it is not much more expensive to purchase a gallon than 3 quarts of paint.

RULE OF THUMB

Choosing the Right Brushes for Wood

For finishing wood with varnish or polyurethane, use an animal-bristle brush or lamb's wool applicator. Nylon bristles tend to leave brush marks and will soften in oil-based solvents. Sponge applicators can also be used for small projects.

To determine the surface area of base and ceiling molding, estimate it at 6 inches wide (regardless of actual width) and multiply this times its length in inches. Divide this number by 144 to arrive at the square-foot surface area.

Interior Wood Finishing

Interior wood-finishing techniques vary, depending on whether you are finishing new wood or refinishing old wood.

Finishing New Wood

If you are remodeling or building new, you will be working with unfinished raw wood. Wood trim and cabinets are presanded at the factory, and require only light sanding to remove any blemishes or scuff marks. Use medium to fine sandpaper and sand the wood lightly. Always sand with the grain. You can use a sheet of sandpaper, folded into thirds and handheld, or attach the sandpaper to a sanding block. For sanding curved surfaces, choose a flexible sanding block that can be pressed to fit the contours of the wood.

If you wish to stain the wood, check the stain samples at your dealer to select the right shade. If you wish to match existing trim, remove a small piece of the trim and take it to the paint dealer so you can compare the stain choices with the stain you wish to match. If no stock stain matches your existing stain, your dealer can custom-mix stain to match your sample. If you must have stain custom-mixed, be sure to buy enough to do the complete job.

Follow the manufacturer's directions for applying the stain. For fast stain application, use a sponge brush, or buy a large sponge and wear rubber gloves to apply the stain. Just pour the stain into a paint tray, dip the sponge into the stain, and wipe it on the wood.

Foolproof Advice
for Staining Wood

If you are remodeling, it is much easier to apply the stain before you install the trim, because you need not protect the walls from the stain. Just set up a pair of sawhorses and lay the trim pieces across the sawhorses. Dip a sponge into the stain, wipe it onto the trim pieces, then wipe the excess away with a clean cloth.

After applying the stain, wipe the wet stain with a clean cloth or towel. The wiping step removes surplus stain and ensures even stain application. Using a sponge applicator for stain as mentioned earlier will minimize the amount of wiping needed.

When the stain has dried, check the shade. If the wood is the right shade, you can proceed with the finishing; if the stain is too light, apply another coat. The shade will also deepen if you let the stain set on the surface longer before you wipe it.

The next step is to lightly sand the stain coat.

Now apply the sealer. Ask your dealer to select the right sealer that is compatible with the finish coat. Let the sealer dry, then sand lightly. Use a tack rag or cloth dipped in mineral spirits to remove the dust from the stained surface. Again, to avoid scratches in the wood, always sand with the grain.

Next, mix wood putty with stain to match the stain you have used. Fill all the nail holes with wood putty. The best application method is to carry a small ball of putty in your hand, and press it into the nail holes with your thumb.

Finally, apply the varnish or polyurethane finish. Choose the right gloss; the finishes are available either in satin or gloss. Again, follow label directions for application.

Refinishing Wood

If you live in an older house and the wood trim or cabinetry looks dingy, don't assume that refinishing is necessary. Often, trim and cabinetry will darken over time, as dust, wax, and grease build up on the wood. Before starting to refinish, clean the existing finish with odorless mineral spirits. This is a good solvent for removing old wax or cooking grease. Wash the wood with the mineral spirits and let dry. Now examine the wood. If the wood is renewed, you have eliminated the job of refinishing it. If you decide to refinish the wood, the cleaning step was not wasted effort, because you must clean the wood thoroughly to remove old wax or grease before refinishing it.

If refinishing is necessary, choose a new finish that is compatible with the old. Newer polyurethanes are the best choice, but if the old varnish, shellac, or paint is in poor condition, you must completely remove it before refinishing with polyurethane. If you will be refinishing over old

RULE OF THUMB
Preserving Wood's Natural Beauty

If you wish to preserve the appearance of the natural wood, apply two coats of polyurethane in either satin or gloss finish. Sand lightly with the grain, and use a tack rag to remove the dust between coats.

varnish, test the polyurethane in an inconspicuous spot before tackling the entire project.

Repainting Interior Wood

To repaint interior wood, first use odorless mineral spirits to remove any dirt, wax, or grease. Then use an alkyd paint for the new finish. Alkyd paints are generally more resistant to wear, smudges, and fingerprints and are easier to clean than latex paints.

Refinishing Floors

If old wood floors are badly scratched or damaged, they must be sanded before refinishing. Hire a pro to resand the floors. If the floors are not damaged but are simply worn, have the floors stripped to remove the old finish.

The best clear finish for wood floors is polyurethane. Polyurethane is durable and resistant to damage from abrasion. For best protection, use two coats of polyurethane on wood floors. You need not use waxes on the polyurethane finish; just damp-mop to remove spills and use an electric broom to clean the floors.

RULE OF THUMB
Staining and Wood Putty

When staining wood it is best to putty the nail holes only after you have applied the sealer. If you apply the putty to unsealed wood, the oil in the putty may leach into the wood around the nail, leaving a dime-size stain around each nail hole.

Paint Strippers/Removers

Paint stripping or removal is a tough task. If the surface has many coats of paint or finish repeated application/scraping/sanding may be required.

A wide variety of chemical paint strippers/removers are available. Your paint dealer can help you select the right product for your particular job. Before deciding to use a stripper/remover, read "What About Lead Paint?" on page 91.

Coverage of paint strippers/removers depends on both the type of finish to be removed and the number of coats to be stripped. Estimate the size of the area you wish to strip, and ask your dealer to recommend the right quantity.

If the finish is basically in good shape with minor peeling, scrape away the loose paint and apply a coat of primer; then, sand and apply the desired top coat over the entire area.

When refinishing over old paint, especially a high-gloss paint or enamel, it is generally advised to first sand the surface. However, sanding creates both a dust mess and possible health hazard from lead paint. For most surfaces, use a liquid paint deglosser product or a strong solution of trisodium phosphate (TSP) to clean and degloss the surface.

Nonchemical Paint Removers

A variety of electric paint-removal tools are available. These include a flat heating tool that can be held against the surface, and heat guns. Because of the danger of fumes from lead paint, use these tools only in a well-ventilated area, and wear an appropriate mask or respirator.

Exterior Painting and Staining

When exterior latex paints were introduced, there were problems of intercoat failure if one used oil-based paints and latex paints in alternate coats. Modern exterior latex paints, however, can be used over any previous paint coating. Because of both their ease of application and cleanup and their low contribution to air pollution, latex paints have become the preferred choice for exterior use.

There are a few exceptions to this rule. Oil-based primers are recommended for the base coat on redwood siding, with latex paint an acceptable top coat; and because of their superior resistance to wear, oil-based stains are recommended for use as the finish coat on decks.

Safety Tips for Using Polyurethane

- Polyurethanes will not adhere to wax. Clean any surface thoroughly before application.

- Use polyurethane finishes in well-ventilated areas.

- Avoid using polyurethane near flames or heat. To avoid fires from spontaneous combustion, dispose of steel wool or rags soaked in polyurethane in a metal pail filled with water, and cover with an air-tight lid.

Tools and Equipment for Exterior Painting and Staining

Oil-based and latex paints and stains can be applied using a wide array of painting tools. Both finishes can be applied using spray equipment, brushes, paint pads, or rollers. The choice of application tools depends to a great degree on the surface to be finished. See "Tools and Equipment for Interior Painting" on page 94 for more information about the following materials.

Spray Equipment

Use paint spray equipment to paint large areas such as the siding on a house, garage, or barn. Spraying is also the best choice if

RULE OF THUMB
Outdoor Spraying

Although spraying is the best approach for painting fences, decks, retaining walls, or house exteriors, be aware that the paint mist can travel or carry a long distance on windy days. This overspray can land on a neighbor's house, car, or other property. Use spray equipment outdoors only on windless days, and only if there is a comfortable distance between your property and your neighbors'. Also check your community's codes before using a paint sprayer; because of property damage from overspray, some communities have banned the exterior use of paint sprayers.

RULE OF THUMB
Buying the Best Exterior Paint

Choose a premium-quality acrylic latex exterior paint, such as Sears Weatherbeater, Olympic Overcoat, or Benjamin Moore paints. The cost of the paint is a small item when compared to the work of painting the house, so don't hesitate to pay a bit more for a quality paint product. Many paints are on sale in the summer, so shop the sales for both quality and low cost, then schedule paint jobs for the autumn, when you can take advantage of cooler temperatures.

the surface is rough or textured, such as painting stucco or other masonry, or for rough-sawn cedar siding. Use a sprayer to paint or stain wood fences with staggered boards or pickets, and for painting window shutters. For painting wrought-iron railings, use an aerosol paint can. Always spray-paint on a day that is not windy, and use plastic drop cloths to control overspray.

Brushes

Brushes for applying alkyd paints should have hog or ox bristles. Brushes with a mix of nylon, polyester, and animal bristles can be used with either latex or alkyd paints. To hold more paint and leave fewer brush marks, choose a brush with flagged tips. Use a tapered sash brush for painting around window sashes. Use a brush for any smooth siding product, including steel, aluminum, hardboard (Masonite), or wood lap siding. Professional painters often use an airless sprayer to apply the paint, and then brush it out.

Brush Size

How-to directions often offer advice on the size of brush to use, but the best advice is to consider your own size and strength first. The basic rule is to use the widest brush possible for wide siding, but a small person will quickly tire when using a brush that is 4 to 6 inches wide. Consider your own stamina when choosing a brush.

Rollers

Buy a quality roller with a fiber core rather than a cardboard core that will quickly fall apart when wet. Use a roller cover that has a ¼-inch nap to apply latex paints; use lamb's wool or mohair rollers for alkyd paints. Rollers are available in either 7- or 9-inch widths. Choose the smaller 7-inch roller for rolling most lap siding.

Drop Cloths

When painting the exterior of a house, use drop cloths to protect landscaping around the perimeter of the house. Remove them as soon as the painting is completed so they don't crush the plantings.

Paint Pads

Use paint pads on the same types of surfaces as you would paintbrushes, that is, on smooth lap siding. Paint pads are especially useful for painting steel or aluminum siding.

Preparation

To prepare any exterior surface for painting, first scrape away any loose or peeling paint. If there are exposed nail heads, drive the nails below the surface of the siding; renail any loose siding.

Caulking

For long life choose a quality acrylic latex or paintable silicone caulk. Check the label on the caulk tube; caulks are now being sold with guarantees of 25 years or more. Use a putty knife to remove old caulk from around window and door frames. Use a heat gun to soften the caulk and make it easier to remove. To prevent water from being forced into the cracks, recaulk before you power-wash the siding. Water that gets into the uncaulked cracks can cause the paint to peel. Depending on the size of the caulk bead, one tube of caulk can fill a crack 15 to 20 feet long. Figure about one tube of caulk for each window and door. When caulking, don't overlook any holes around pipes or wires, or at any point where dissimilar materials meet, such as where siding meets a brick fireplace chimney.

Painting Metal Siding

Steel and aluminum siding is prepainted at the factory and usually does not require repainting for at least 20 years. When the time comes to repaint steel or aluminum siding, first clean the siding with a good deck cleaner to remove oxidized paint. Then paint with a quality acrylic latex paint. Check the label to be sure the paint you choose is approved for use on aluminum.

Power Washing

To clean any large surface for painting, rent or buy a power washer. Rental power washers may cost about $65 per day, but you can easily clean your house, gutters, walk, deck, drive, and garage floors in a single day. The high-pressure washers will blast away grime and mildew that will not yield to ordinary garden-hose water pressure. Remember, the single greatest cause of paint failure is poor preparation of the surface, so take time to be sure the surface is truly clean before applying the paint.

If the siding is really dirty from grime or mildew, wet the surface to lift the grime, then wash with a 50-50 mixture of clean water and chlorine bleach. Let set until the bleach loosens the grime, then power-rinse with the washer.

Troubleshooting Exterior Paint Problems

Before repainting your house, look for telltale signs of paint problems. Modern paints are chemically formulated under controlled conditions, so it is rare when a paint problem is actually due to paint quality. When alligatoring, blistering, checking, efflorescence, fading, mildew, or peeling of the paint occurs, suspect one of these underlying conditions, and take steps to eliminate the cause before repainting (see figure 6.1).

Alligatoring

Alligatoring occurs when paint develops a scalelike or "alligator skin" effect. Alligatoring can be caused by excessive expansion/contraction of the siding, by applying a noncompatible coating over an existing paint coat, or by a thick buildup caused by too many coats of paint.

Cure. Remove the old paint by scraping, sanding, or using a heat gun. Apply a primer and a top coat.

Blistering

If paint blisters, suspect that either the paint was applied on an unclean surface or there is moisture penetration behind the siding. If there is no vapor barrier in outside walls, interior moisture can migrate through the walls until it reaches the paint film. When the sunlight reaches the siding, the warmth will cause the moisture to lift the paint and produce a paint blister.

Cure. Paint will not stick to mildew or dirt. Always be sure the surface to be painted is clean. Scrape away the loose or blistered paint, then prime and repaint. If you have no vapor barrier in exterior walls, apply a coat of alkyd paint to the interior surface of the exterior walls. Use caulk to seal any cracks between the wallboard or plaster and the electrical outlets.

Checking

Both the causes and cure for small ladder-like cracks, or checking, are similar to those described for alligatoring. Follow the instructions given there.

Efflorescence

Efflorescence is most often seen on stucco or other masonry materials. It is a white powdery substance caused by salts in the cement leaching outward to the surface of the masonry.

Cure. Let masonry construction age for 6 months or more before painting. Use a wire brush to remove efflorescence, then repaint. Use only masonry paints, not house paints, for painting masonry surfaces.

Fading

This is one problem that may be caused by failure of the paint pigment. It is often observed on surfaces that are painted dark blue or brown. If the fading is extensive over the entire house, complain to the

Figure 6-1 Paint Problems

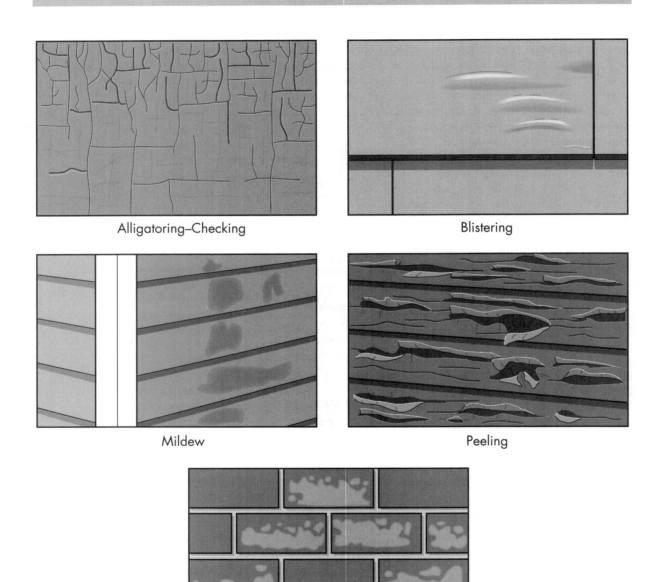

Alligatoring–Checking

Blistering

Mildew

Peeling

Efflorescence

dealer, who may make an adjustment. However, if the paint on only one side (usually south or west) of the house is faded, suspect too much direct sunlight.

Cure. Repaint. When buying paint, query the paint dealer regarding the fade resistance of the paint, or call the paint manufacturer via an 800 number. (A directory of 800 numbers is available at public libraries, or call 1-800-555-1212.)

Mildew

Mildew can develop on any surface that is subject to heat and moisture. Power-wash your house each spring to remove grime that will support mildew growth.

Cure. If mildew is already present, mix chlorine bleach and water in a 50-50 solution. Use a garden sprayer or power washer to apply the bleach mix on the siding. Let set until the bleach lifts the mildew, then wash thoroughly with clear water. Surfaces that are shaded do not dry thoroughly. Trim shrubbery or overhanging tree limbs that shade the siding so sunlight and air can dry the surface.

Peeling

By a wide margin, peeling paint means poor adhesion, and poor adhesion is almost always due to painting over a surface that was improperly cleaned and prepared. As with blistering, suspect that when paint peels, paint was applied to a dirty surface.

Cure. Always be careful to clean and prepare the surface properly. If peeling is occurring only at siding joints or where siding meets door or window jambs, scrape away peeling paint; then, prime, recaulk any cracks, and repaint. If paint is peeling on siding on the house's gable ends only, the cause is insufficient attic ventilation. Add more gable vents, or install an electric vent fan to remove attic moisture.

7 Walls

The walls of a house make up about half the total interior surface, so the way walls are finished can have a very dramatic impact on the house interior. Exterior walls provide weatherproofing and shelter from the elements. Interior walls or partitions delineate space, and provide privacy screens, fireproofing, and soundproofing.

If you are redecorating walls, study carefully the impact that color, texture, and lighting can have on a room. Changing the color of the walls in a room can completely transform the appearance and mood of the room. Direct or side lighting will highlight any cracks, nail pops, bulges, or other defects on a wall, while indirect lighting can make any wall appear smooth and flat.

Plaster

Until the post–World War II building boom, plaster was the common interior wall and ceiling finish. Plaster consists of a base of lath, wood strips nailed to the wall studs, followed by the first coat of plaster, called a brown or leveling coat, which is then troweled smooth with a thin finish coat of white plaster or lime. When damage occurs to plaster, the failure is usually between two or more of these layers.

Troubleshooting Plaster Problems

To determine the condition of plaster walls, first visually inspect the surfaces for cracks, holes, or bulges. Cracks and bulges may indicate a failure of the plaster layer, meaning that the plaster may be loose from the lath itself. The severity of this problem depends on how extensive the area of loose plaster is. Small areas may mean that the brown coat may be loose from the lath, or there has been a separation of the finish coat from the brown coat.

To determine the area of loose plaster, press with the flat of your hand on any bulges and on both sides of any cracks. If the plaster feels loose or spongy under hand pressure, that plaster is loose and must be removed and the area replastered. If the spongy feeling extends over a wide area, you may have to remove the plaster from the entire wall and replaster it, or install wallboard to replace the plaster.

Cracks

Cracks may be of the hairline variety, that is, minor cracks that are caused by expansion and contraction of the framing and/or the plaster. These cracks can readily be repaired. Because these minor cracks are due to expansion, which occurs each time the seasons (i.e., temperature and humidity) change, the plaster will be subjected to these forces as long as the house stands. Therefore, any superficial approach, such as applying a patching material called spackling or patching the crack by plastering, will result in failure, because the cracks will reopen the next time expansion occurs.

To permanently cure hairline cracks, tape and finish the cracks exactly as you would tape and finish the cracks or joints in wallboard. To make this patch, first you have to remove any loose plaster. Then, use a wide 6-inch putty knife, called a taping knife, available at home centers and paint and hardware stores, to coat the cracked area with ready-mixed, all-purpose taping compound. Cover only a small area to avoid having the compound dry before the tape is placed. Then use paper (not fiberglass) tape to cover the cracks. Using the taping knife, press the tape firmly into the compound and remove any excess compound. Proceed until you have covered all cracks, then let the compound dry before applying the covering coats of taping compound.

Taping compound, also known as joint or finish compound, is available either as a dry powder to be mixed with water, or ready-mixed in pails. Because the consistency of the taping compound is critical to the success of the job, and to avoid the dust and mess of mixing, always use ready-mixed compound for repair and remodeling jobs.

When the tape coat is dry, use a 10- or 12-inch taping knife to apply a thin second coat. This coat is intended to smooth and conceal the taped area. Let the second coat dry completely, then use the wide taping knife to apply a third and final, finish coat.

Foolproof Advice for Avoiding Plaster Dust

To avoid plaster dust, use a wet sponge to smooth the compound over the taped cracks. Keep a pail of water handy for cleaning and rewetting the sponge. Special sponge wallboard sanders are available for this purpose, or an ordinary large household sponge will do the job. Sponge-sand the finished cracks as soon as the compound has hardened but is not completely dry.

Holes

To fill and repair small holes in plaster, use low-shrink patch plaster, which shrinks less than taping compound. Because patch plaster is a chemical-set material, as opposed to the air-dry taping compound, patch plaster cannot be shipped premixed. Mix only as much plaster as you can use in 1 hour, before the plaster sets in the pail.

To patch a small hole, first clean all loose plaster and dust from the hole. Use an old paintbrush or an inexpensive disposable paintbrush to paint a coat of latex concrete/plaster bonding liquid, available at home centers, on the lath and edges of the hole. Then mix the patching plaster to a workable consistency, that is, stiff enough to

RULE OF THUMB

Mixing Patch Plaster

Patch plaster will harden in the pail, or other container, making the pail hard to clean. Always use a plastic pail for mixing patch plaster. When the material hardens in the pail, simply flex the sides of the pail to pop the hardened material out of the pail. Dispose of the hardened plaster by placing it in your trash container.

hold on the trowel but soft enough to smooth and level easily. Apply the patch plaster to the hole, pressing it firmly into place so that the plaster is forced between the cracks in the wood lath, and grips the edges of the surrounding plaster. If the hole is more than ½-inch deep, apply the plaster in two coats. First fill the hole to more than half its depth, then use a stiff brush or the edge of the trowel to create ridges or striations on the patch plaster surface. This step will help the second coat of plaster bond to the first coat. When the plaster is hardened, apply and smooth a second coat over the first coat.

If the hole in the plaster is larger than a football, or there is loose plaster over a large area, it requires a degree of skill to patch the hole. The best advice is to call in a professional.

If you have a hole that is grapefruit-size or smaller, you can buy a peel-and-stick patch at your home center or paint store. Simply remove the dust and loose

plaster from the area, peel the backing off the patch, and press the patch in place over the hole. Then use taping compound to smooth and conceal the press-in-place patch. Sand or sponge-sand lightly, then apply a coat of wall primer over the patched area.

Wallboard

During the emergency housing boom of World War II, gypsum wallboard became the interior finish of choice. Wallboard installation and finishing require less time to complete than conventional plaster. However, new thin-coat plaster eliminates the drying time required for the thick brown coat, so it also provides a time advantage.

Wallboard panels are available in lengths from 8 to 12 feet, and in thicknesses of ¼ inch (available on special order), ⅜, ½, ⅝, and ¾ inch.

RULE OF THUMB

Wallboard Panel Thickness

Never use ⅜-inch wallboard panels as a single layer for wall or ceiling applications. This thinner panel does not have the impact resistance necessary for wall application, and when used for ceilings the thin material will sag between the joists, especially if blown or loose-fill insulation will be installed above the ⅜-inch panels. Use ⅜-inch wallboard as a base or substratum under prefinished wall paneling, or in double layers, called laminated wallboard.

Laminated Wallboard

Double-layer wallboard, called laminated wallboard, calls for two layers of ⅜-inch wallboard, with the first layer nailed or screwed to the framing, and the second layer glued over the first. This system provides a ¾-inch-thick wall or ceiling equal in mass to plaster, and therefore provides superior soundproofing and fireproofing to single-layer, ½-inch-thick panels.

Because the second layer of wallboard is glued to the first, the second layer is installed using few if any nails or screws, thus eliminating not only the need for treating the fasteners with wallboard compound to conceal them, but also the possibility of nail or screw pops, a common complaint with single-layer wallboard construction.

For laminated wallboard, the first or base layer of panels is installed so that the panels lay parallel to the ceiling framing, with the second ceiling layer run perpendicular to the framing. Joints are staggered so that the joints do not occur in the same location on both layers.

For wall application, stand 8-foot-long wallboard panels up so the joints occur parallel to the studs. Then install 12-foot-long panels perpendicular to the wall studs, again being careful that the joints are staggered between the two layers.

Use contact adhesive or wallboard taping compound to glue the two layers together. Use a notched adhesive trowel to spread the adhesive on the first layer of wallboard. For ceiling application you must apply screws at the edges of the panels, and one screw at the center of the panels, to hold them in place until the adhesive sets. The tape and finish will cover the screws at the panel edges; there will be a single screw at midpanel, on each ceiling joist. These screws must be covered with three coats of compound.

For wall application, only temporary nailing is required. This can be done using double-headed form nails, driven through holes in wood lath. The lath is temporary and, along with the nails, is removed after the adhesive is set, leaving no nails to be treated in the center portion, or field, of the panel. Screws applied at panel edges can be covered with tape and compound, and need not be removed.

Special Wallboard Panels

In addition to the familiar plaster-core wallboard, other wallboard panels are offered for special applications. An example is water-resistant panels, called green board because of their color, that are used in areas where water or humidity problems may be encountered, such as bath or laundry rooms. Green board was once the preferred product for use as a substratum or base for tile application around bath and shower areas. Today, a cement-based panel called Durock is preferred for use as a base for ceramic tile. Figure 7.1 shows an installation of Durock around a bathtub.

RULE OF THUMB

Using Green Board

Because they may sag, water-resistant or green-board wallboard panels should never be used for ceilings.

Do not install plastic vapor barriers over wall framing where you will install green board.

Figure 7-1   Installation of Durock around a Bathtub

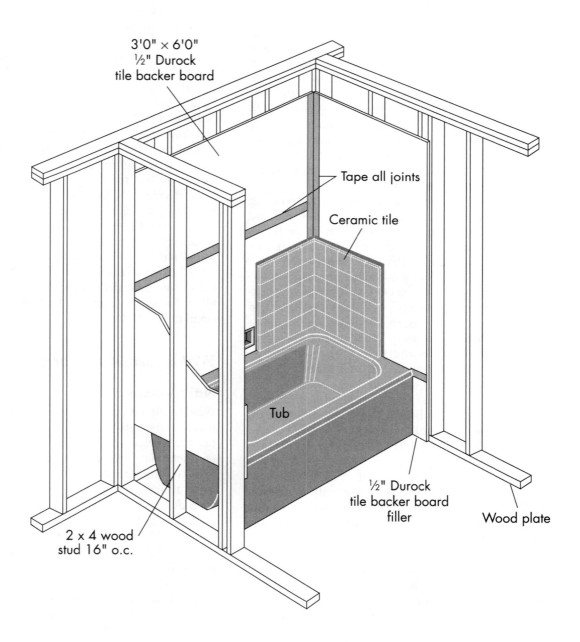

3'0" × 6'0"
½" Durock
tile backer board

Tape all joints

Ceramic tile

Tub

½" Durock
tile backer board
filler

Wood plate

2 x 4 wood
stud 16" o.c.

The disadvantages of wallboard versus plaster are that wallboard provides less thickness or mass in the wall, and so lacks the superior soundproofing and fireproofing qualities of conventional ¾-inch-thick plaster. However, the familiar ½-inch wallboard is adequate for most residential uses, and ⅝-inch wallboard provides a 1-hour fire rating when applied on the firewall between the house and an attached garage. Because of the wider spacing of roof trusses (24 inches o.c.), and because modern energy standards require more ceiling insulation, which increases the weight on the ceiling, the thicker ⅝-inch wallboard is also recommended for ceilings.

If better soundproofing or fireproofing is desired, two layers of ⅜-inch wallboard can be applied to provide the same thickness or mass as conventional plaster (see "Laminated Wallboard" on page 116).

To increase the sound-absorbing qualities of a wall, install sound-absorbing insulation between the wall studs. The insulation will help absorb sound waves traveling through the wall. While sound-absorbing insulation is effective, some sound energy will pass through the wallboard stud sandwich. See figure 7.2.

A more effective sound-deadening wall system can be constructed by using a 2 × 6 for a top and bottom plate instead of the standard 2 × 4. By installing the 2 × 4 studs flush with the edge of the 2 × 6 plate the studs only come in contact with the drywall panels on one side of the wall. Insulation placed between the stud rows make this type construction a very effective sound-deadening wall.

Other systems are in common use. By installing resilient metal furring channels to one side of the studs, the drywall panels are isolated from one another. These furring channels will not transmit sound, and will increase the effectiveness of the wall structure in absorbing sound. This system is often used to construct ceilings.

Estimating Wallboard

The most accurate way to estimate wallboard is to measure all the wall and ceiling areas to be covered. Suppose you will apply wallboard to all wall and ceiling surfaces in a room that is 20 by 12 feet. The ceiling area is thus 240 square feet. The wall height is 8 feet, so multiply 40 feet (two 20-foot walls) times 8 feet, which equals 320 square feet. Multiply 24 feet (two 12-foot walls) times 8 feet, which equals 192 square feet. Your total thus will be 752 square feet of wallboard. Subtract from this the area for any large openings; for example, subtract 21 square feet for each 3-by-7-foot door. Now divide the total remaining square footage by the number of square feet in the size of panel you will use; for example, divide by 32 square feet if you're using 4 × 8 panels, or 48 square feet if using 4 × 12 panels. This will give you the number of panels needed for the job. Be sure to allow for waste from cutoffs that are too small to be usable.

Wallboard Application

Nails versus Screws

When wallboard was first introduced, there was a serious problem: no special nails or fasteners were available for applying the product. Thus, carpenters used ordinary 5d nails, or large-headed galvanized roofing nails, to fasten wallboard. (See chapter 2 for a discussion of nails and other fasteners.) This lack of a special fastener led to many problems, chief among these being fastener failures or "nail pops." Nail pops result when a nail is driven into moisture-laden lumber, because the lumber will shrink as the house settles. The degree or severity of the nail pop will be directly proportional to the length of penetration of the fastener and the amount of moisture in the framing member. Eventually special

Figure 7-2 Soundproofing a Wall

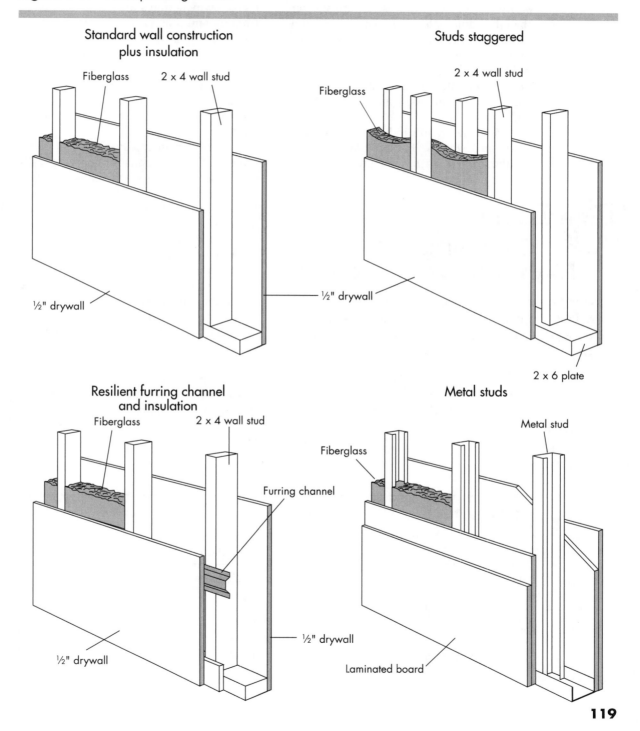

Standard wall construction
plus insulation

Fiberglass   2 x 4 wall stud

½" drywall

Studs staggered

2 x 4 wall stud

Fiberglass

½" drywall

2 x 6 plate

Resilient furring channel
and insulation

Fiberglass   2 x 4 wall stud

Furring channel

½" drywall

½" drywall

Metal studs

Metal stud

Fiberglass

½" drywall

Laminated board

nails were developed, called ring-shank wallboard nails. These nails were shorter than the commonly used 5d nails ($1\frac{1}{4}$ inch versus $1\frac{1}{2}$ inch), but because of the annular rings on the nail shank they had superior holding power, and because there was less penetration into the framing they reduced the degree of nail pop. But nail pops were still a problem.

Much later, the industry introduced wallboard screws. For applying $\frac{1}{2}$-inch wallboard, 1-inch screws are used. The screw shank offers good holding power, but the screw penetrates only $\frac{1}{2}$ inch into the framing and thus virtually eliminates the problem of nail pops.

Because fewer screws than nails are required for applying wallboard, there are fewer fasteners to treat, reducing finishing work. Table 7.1 shows the maximum spacing for the most common size drywall screws.

There are other advantages in using screws rather than nails for applying wallboard. Because nails are driven with a hammer, there can be damage to the wallboard from the hammer impact, especially around the nail heads. This damage may be half-moon–shaped paper fractures around the head, which will show through the wall paint as tiny cracks. Or the impact may be too severe and may crush the plaster core under the nail head, thus destroying the holding power of the nail.

Also, when the nail is set below the surface of the wallboard, you cannot control the depth of the indentation from the hammer head. Variations in framing lumber density will cause some nails to be set only slightly, while others are deeply set. This gives uneven results when you apply the three coats of compound needed to fill the

RULE OF THUMB

Choosing the Right Screw or Nail

Using a fastener that is longer than necessary defeats the purpose of the improved fasteners: sufficient holding power with a minimum of penetration into the framing. Use only $1\frac{1}{4}$-inch-long ring-shank nails, or 1-inch-long wallboard screws, for applying $\frac{1}{2}$-inch wallboard. Use fasteners that are $\frac{1}{4}$ inch longer for applying thicker $\frac{5}{8}$-inch wallboard.

Table 7.1 Drywall Fastener Spacing

Fastener	Application	Maximum Spacing
Drywall nails	Ceilings	7"
	Walls	8"
Drywall screws: framing on 16" centers	Ceilings	16"
	Walls	12"
Drywall screws: framing on 24" centers	Ceilings	12"
	Walls	12"

indentations. By contrast, the wallboard screw gun has an adjustment ring that permits you to release the screw at a predetermined depth, so all screws are automatically indented at the same depth.

For all the above reasons, one should always choose screw application rather than nailing wallboard. Wallboard screw guns can be purchased for around $50, rented, or perhaps borrowed.

Wallboard Adhesives

Wallboard adhesives are available in tubes that fit your caulk gun. Using adhesives in conjunction with wallboard screws permits you to achieve the best possible wallboard construction, with a minimum of fasteners to fail or treat. Adhesives are applied as a ¼-inch bead (line) along the stud or joist. Using adhesives will also help correct small problems with framing alignment, so you will have a flatter, smoother wall if you use adhesives. Note that you obviously cannot use adhesives on walls that are covered with a plastic vapor barrier.

Adhesives are especially useful in problem applications. Use adhesives to apply thin strips of wood called furring strips over concrete walls, then to apply the wallboard over the furring strips. If you are applying wallboard over a pocket door, use adhesives to avoid driving screws or nails completely through the ¾-inch-thick pocket framing. Use adhesives at any area where you must apply wallboard over steel beams or concrete surfaces.

For superior soundproofing, use Acoustical Sealant. This product can be used to prevent sound transmission through cracks around electrical outlets, pipes, doors, or at wall and/or ceiling joints. Ask your dealer to order the sealant for you, or call wallboard suppliers that serve professional installers.

Wallboard Tools

Tools needed to install wallboard include a screw gun, hammer, 12-foot measuring tape, carpenter's pencil or crayon, razor knife, keyhole or saber saw, handsaw, 4-foot-long wallboard T-square, and a wallboard lifter for walls to lift the bottom wallboard panel up against the top panel.

Wallboard Layout

Because getting the best results requires using the longest panels to reduce the number of joints to treat, always use the longest panels possible when installing wallboard. The standard used by professionals is 12-foot-long panels. These long panels are both heavy and unwieldy, so at least two sturdy workers are needed to handle the material. Do not use shorter, 8-foot panels simply because they are easier to handle; this will result in having extra joints, meaning additional work in finishing the wallboard. The single exception to this advice is when hanging wallboard where room access will not permit carrying in the longer, 12-foot panels, for example, when finishing an attic or basement where stair configuration does not permit using the longer panels. If you cannot handle the longer panels, consider hiring professionals to hang the wallboard.

Vertical versus Horizontal Application

Many how-to texts do not discourage using 8-foot-long panels, installed vertically (with joints parallel to the studs) for wallboard installations on walls. However, all wallboard manufacturers recommend horizontal application of wallboard panels for all 8-foot-high walls, and we strongly recommend horizontal application.

The reasons for advising horizontal application are many. First, using 12-foot panels, installed horizontally or perpendicular to the studs, will result in less footage of joints to treat. Fewer joints means a flatter, smoother wall, with less chance of joint cracks. Also, joints that occur at 4-foot intervals and extend from floor to ceiling are difficult to conceal, because they will all be at eye level regardless of the height of the occupants of the house. Also, if the stud is crooked, the joint that occurs on it will be distorted and will result in a joint over a bulge in the wall. Figure 7.3 shows horizontal and vertical drywall applications for walls and ceilings.

Horizontal application also means that most of the joints on the wall occur near waist level for more comfortable joint finishing. Wallboard panels will crack easily along their length, but are stronger across the width of the panel. The horizontal panels also tie together a maximum of framing studs, so the wall will be stronger if the wallboard is installed horizontally.

Wallboard Layout Plan

Use graph paper to lay out the wallboard installation. Plan no joints at points where the framing changes direction, such as at edges of doors or windows. Instead, let the wallboard panel be continuous over the opening, and then cut out the wallboard in the opening area. This will eliminate having joints at corners of doors or windows, where the joints are certain to crack.

If the wall is so long that you must have end-to-end, or butt, joints, lay out the installation so butt joints occur as near as possible to the wall ends or corners, not in the middle of the wall. Do not plan joints so they occur on the same stud as plumbing pipes or electrical boxes, because having these interruptions on a joint makes finishing difficult.

Checking the Framing

While irregular framing is corrected in a plaster wall by applying a thicker coat of plaster where necessary to straighten the wall, wallboard is uniform in thickness, so having a straight wall means that framing must be straight. Use a 6-foot-long (or longer) straight board to check the framing stud alignment. Place the straightedge perpendicular to the studs and check along its length for any studs that are bowed or misaligned. If the stud is merely misaligned, you can usually pound it with a hammer to align it with other studs. If the stud is bowed, remove it and replace it with a straight stud. In this manner straighten the framing over the entire wall or walls to be covered.

When checking the framing, look for any corners where there is no backing to support the wallboard panel ends. Install backing or nailing strips where necessary. If you find that the wood bracing used to keep the joists aligned, called bridging, protrudes beyond the joist face, cut off the ends of the bridging. You may find it easier to replace the wood bridging with metal joist bridging, which takes up less room and thus will not interfere with wallboard application.

RULE OF THUMB

The Price of an Odd-Angle Wall

An odd-angle wall will cost twice as much as a wall built square.

Figure 7-3 Wallboard Layout

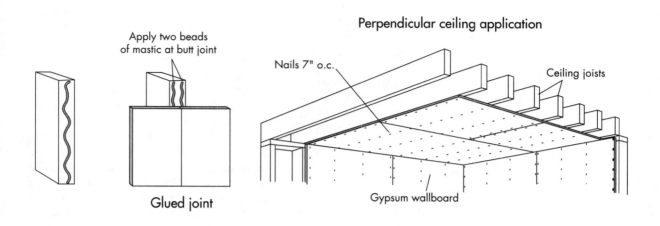

Apply two beads
of mastic at butt joint

Glued joint

Perpendicular ceiling application

Nails 7" o.c.

Ceiling joists

Gypsum wallboard

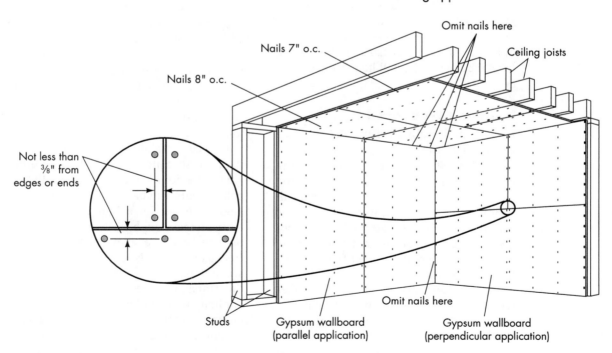

Parallel ceiling application

Omit nails here

Nails 7" o.c.

Ceiling joists

Nails 8" o.c.

Not less than
3/8" from
edges or ends

Studs

Gypsum wallboard
(parallel application)

Omit nails here

Gypsum wallboard
(perpendicular application)

Hanging Wallboard

When hanging wallboard, the goal should be to create as smooth and straight a wall as possible. This means using only the minimum number of fasteners recommended, using the longest wallboard panels possible, and planning to eliminate unnecessary joints. Wallboard is inexpensive, so you should not try to use up small pieces of scrap wallboard to save a few dollars. It is better to install the wallboard so that you have a minimum of joints, and dispose of the scrap.

Cutting Wallboard

Wallboard is delivered with two panels taped together, with the back (or gray side) of each panel facing outward. When stocking the wallboard in the room, or before cutting the wallboard, pull off the end tapes to separate the panels. Then place them standing up against a wall so that all the face (or finish) sides of the panels face outward. Placing all panels so the finish side is out, and therefore measuring and cutting from the finish side, will help avoid cutting errors.

Installing Wallboard

Because most walls are usually up to 1 inch higher than 8 feet (the combined total of two 4-foot-wide panels), when hanging wallboard on walls, always hang the top panel first, then the bottom panel. This will result in a tight-fitting joint between the ceiling and wall, with the crack at the floor level where it can be covered by base trim. If you hang the bottom sheet first, and set the top sheet above it, then you will leave a crack at the ceiling-wall joint. Older homes with high ceilings should have the top panels installed first, then the bottom and a strip of drywall cut to fit between the panels. This way the joints will be at waist height and easy to tape.

When installing wallboard on walls, use a carpenter's pencil to mark the stud locations on the floor. This will help you find the studs for driving the screws when the wallboard panels are placed and you cannot see the studs. Also mark the locations of electrical boxes on the floor, to avoid covering them over.

Measure the length of the wall. If the wall length is less than 12 feet, subtract ¼ inch to allow the panel to fit easily between the corners on the wall. If there is a window in the wall, cut out the wallboard for the window opening. For walls with doors, hang the top panel, then cut out the opening using the door framing as a guide. Use a razor knife, with a 4-foot T-square as a guide, to cut the wallboard panels for length. Cut through the face paper, bend the cut end of the wallboard back to break the paper core, then cut through the back paper to separate the two pieces.

If the wall is longer than the 12-foot-long panel, carefully select the stud on which the end or butt joint will fall. You are not required to place the butt joint on a particular stud; for example, if you measure from one wall and the stud nearest the 12-foot mark is crooked, then measure from the opposite wall and place the butt joint at the other end of the wall.

Place a straightedge so it extends over the joint stud plus the two studs on either side. The joint stud should be flat, or better, slightly in back of the two adjoining studs. Do not let a butt joint occur on any stud that is high, such as one that sticks out beyond the two adjoining studs. Because the ends of wallboard panels are not tapered, as the long edges are, the finishing tape and compound when applied will make the finished joint protrude out from the wall, creating a hump or bulge at the joint location. For this reason, the stud that a butt joint falls on must be at least flat or straight with the adjoining studs in order to build a flat joint.

Measure from the corner or end of the wall to the center of the stud where the joint will be. If you are hanging a top panel, start nails or screws in the top edge of the panel while it is sitting upright on the floor, against the wall. Start the nails ⅜ inch from the edge and align them so that, although they will be driven into the top wall plate, they are in line with the studs. Then lift the top panel into place so it fits snugly against the ceiling, and beginning in the center of the panel, drive the started fasteners home as per the manufacturer's directions.

Carefully measure and mark all electrical outlets, then use a wallboard keyhole saw or saber saw to cut out the openings. When cutting openings for wall outlets that occur in the lower panel, measure the electrical box from the lower edge of the top panel, down to the top and bottom of the electrical box. This is necessary because, if you measure from the floor up to the box, the opening will be wrong when you lift the panel into place against the upper panel.

When hanging wallboard, hang one wall completely, both upper and lower panels, then move to the adjoining wall. If you prefer, you can drive in only enough screws to hold the panel from falling, then go back and drive all remaining screws at one time.

Finishing Wallboard

Materials needed to finish wallboard include fiber-paper joint tape, premixed taping compound, corner bead (as required), and wallboard nails to secure the bead. Tools needed include 4-, 6-, 10-, and 12-inch taping knives, a mud pan to carry compound to the wall, a tin snip to cut corner bead, a hammer to nail the beads, and a sanding block or wallboard wet sander.

Wallboard finishing should be done in three steps, or coats.

- When taping, use a 6-inch-wide taping knife to apply taping compound to all butt joints where the untapered edges of the drywall panels meet. Then embed the tape into the compound. Next apply compound to the joints formed by the tapered edges of the drywall panels and embed tape into this compound. Finally, apply compound and tape to the corners.

- To reduce sanding, wipe away excess taping compound from under the tape, and clean off spilled compound as you go.

- After taping, nail on all metal or paper-and-metal corner protectors, called corner bead. The corner bead is installed over the edges of the drywall to prevent damage. Use a 10-inch taping knife or trowel to apply the first fill coat over the corner bead. Coat or spot all screw heads with drywall compound as you did on the ceiling and walls. Allow this first coat to dry for 24 hours.

- On the second finish coat of drywall compound, use a 10-inch taping knife to apply compound over the butt or end joints. Try to keep the finish coats flat—avoid piling compound over the joints. Then coat the long or recessed joints. Again, the goal is to end up with a joint that is flat, so do not pile compound over the joint. You are merely trying to fill the recessed edges so they are flat.

- Using the same knife, apply a second coat of compound over the corner bead. Use a 6-inch taping knife to apply a second coat over the screw heads.

- To finish the inside corners, use a 4-inch taping knife to smooth one side of the corner. Let this second coat dry.

● Proceed as above to apply the third and final coat. Use a 12-inch taping knife and feather the edges of all joints and corner bead out beyond the second coat. Apply the third coat over screw heads, and use a 4-inch knife to finish the remaining side of the inside corners. Let dry.

If you have worked carefully, keeping the treated areas free of compound spills and smoothing all edges, the sanding job will be easier. The easiest and most dust-free sanding method is to use a household sponge or one designed for this job, called a sponge wallboard wet sander, to smooth the job. This is best done after the compound has begun to set up firm, but before it is totally dry. At this point the wet sander will easily smooth the compound.

Decorating Wallboard

Decorating wallboard presents a special problem because you are treating two different materials: the face paper and the taping compound. Never use alkyd or oil primer on wallboard that will be painted, because it will cause the nap on the paper to rise and present a rough surface for decoration. Use an alkyd primer only on ceilings that will be spray-textured.

The best primer for wallboard is a special product called Sheetrock First Coat. This product will equalize the absorbency over the entire wallboard surface, so the face paper and taping compound will absorb paint equally. This product will minimize or eliminate the problem known as "joint banding," which means that the areas treated with joint compound will appear smooth when primed, while the paper surface has a slight texture. This occurs when ordinary paint is used as a primer coat on wallboard.

Let the Sheetrock First Coat dry, then decorate with paint or wallpaper as desired. Be aware when painting wallboard that acrylic latex paints contain solids that will also help conceal the joint compound treatment and will make the walls look smooth.

Paint "hideability" is highest with latex paints. The sheen or gloss of the paint will also affect the appearance of the walls. Flat paint with no sheen conceals blemishes best, then semigloss, then high-gloss paint. Any paint that has sheen—that is, reflects light—will emphasize the blemishes or defects on any painted surface.

Wallboard Repairs

Holes

For patching small holes in wallboard, up to 4 inches in diameter, use a peel-and-stick adhesive patch, available at paint departments. Simply pull the backing strip away from the adhesive and position the patch over the hole. Then apply a very thin coat of compound over the patch and let it dry. When dry, apply a second smoothing coat of compound. When dry, sand lightly and apply primer, then paint.

To repair larger holes, you must cut out the damaged area and install a new piece of wallboard. Use a carpenter's square to mark off a square or rectangular area that overlaps the damaged area, and extends to the middle of the studs on either side of the hole. Then use a saber saw to cut out the damaged area. Over the studs, where you cannot make the cut with a saber saw, make repeated cuts at the stud centers with a razor knife to cut completely through the wallboard.

Now cut a wallboard patch to fit the hole. Insert the patch into the hole and secure it on both sides with screws driven into the studs. Tape the cracks on all four sides of the patch and apply two coats of taping compound over the tape, allowing each coat to dry completely before applying the next coat. Sand lightly and prime or paint.

Nail Pops

As mentioned earlier, use wallboard screws and dry framing lumber to minimize nail or screw pops. If you already have nail pops on existing walls, drive a wallboard screw about 2 inches from the popped nail, then use a hammer to drive the popped nail down so it is dimpled below the surface of the wallboard. Do this to all popped nails. Next, use a 6-inch taping knife to apply taping compound over the repairs. Apply three coats, allowing drying time between each coat. Sand lightly and apply two coats of paint. Because the framing lumber in existing walls is dry, nail pops will not recur.

Don't use a nail set to redrive popped nails. The nail set will fracture the face paper around the nail head if you drive it completely through the wallboard, and when painted over, the paper fracture will appear as a half-moon–shaped crack around the nail head.

Cracked Joints

When wallboard is properly installed and finished, joint cracks are rare. Usually, any such cracks are structural cracks, caused by movement of the framing when temperatures or humidity levels change. Because there is perpetual movement in framing, such joints are likely to recrack after being repaired. The repair procedure is to sand over the taped joint to remove as much compound as possible. Then coat the crack with compound and tape it. Apply two very thin coats of compound to conceal the tape.

If careless workers apply a too-thick coat of compound or paint over a corner, a fine crack may develop. Often, this crack is through the thick paint or compound only, not through the tape. To repair these corner cracks, fold a piece of sandpaper in half and, using the folded edge, sand away the compound buildup at the corner. Then apply a thin coat of paint.

Paneling

Solid planks of cherry, oak, or walnut were once used for paneling, but because of their cost they were usually used only in the most expensive houses. The development of plywood paneling made paneling affordable for all homeowners.

For many years, low-cost plywood paneling was used as a quick cover-up for deteriorating plaster walls. That fad soon passed and plywood paneling lost its popularity. Today, paneling has become popular for use as an accent for family rooms, home offices, and dens, and for wainscoting in kitchens or dining rooms. (Wainscoting is paneling that is applied only on the lower portion of a wall, usually to a height of 32 or 48 inches.)

The pluses of plywood paneling include low maintenance, freedom from redecorating chores, durability, and ease of cleaning. On the downside, having plywood paneling means you are stuck with the color and effect of the paneling. It cannot be painted.

Selecting Paneling

On the theory that you don't always get what you pay for, but you never get more than you pay for, it is wise to avoid cheap or bargain-priced paneling. Avoid hardboard products that have a vinyl face that simulates real wood. Select solid plywood paneling that has a face ply of genuine wood.

Estimating Paneling

Because you cannot have unnecessary joints in a paneled wall, you cannot use all the panel scraps. Thus, much of the cutoff material will be waste. For this reason, when estimating paneling you must figure for solid walls, as though there were no openings. (Subtract only the square footage for very large openings, such as a picture window or double doors.) If you will be paneling a room that has an 8-foot ceiling, each panel will reach from floor to ceiling. Just measure the lengths of the four walls of the room and divide the total by 4 feet, the width of the panels. Thus, a room that is 20 by 12 feet will total 64 running feet. Divide 64 by 4 and you will need 16 sheets of paneling. Some paneling is available in longer sheets to accommodate higher ceilings. Ceiling and wall molding can also be used to cover gaps at the top and bottom between the paneling and the ceiling or floor.

Plywood Panel Sizes

Standard plywood paneling is offered in 4-by-8-foot sheets. Thicknesses include ¼, ⁵⁄₁₆, and ⅜ inch. Panels of these thicknesses are intended to be applied over existing walls, or for new construction, over a base or substratum of ⅜-inch wallboard.

Tools Needed

Plywood paneling can be cut with a power saw, handsaw, or saber saw. Other tools needed include a measuring tape, hammer,

Cutting Paneling

To avoid splintering along the cut line, use a plywood blade in your circular saw, and cut from the back side. This way, the rotation of the blade teeth will be upward into the finished side. When cutting paneling with a table saw, place the panel on the saw table with the finished side up.

nails, panel adhesive, caulk gun, small wood plane, pry bar or lifter to lift the panels into place, carpenter's pencil, sandpaper (to smooth cut edges), square, carpenter's level, chalk box, and protective eye goggles.

Transporting 4 × 8 Paneling

As awkward as 4 × 8 panels are, you can transport them to the job by gripping them along the side edges and hoisting them on your back. With the material standing on end, position yourself with your back to it and grip the two sides by reaching behind you. Lift the panel and hoist it on your back as you walk forward.

To transport paneling on a car or vehicle, use a roof rack and two parallel 2 × 4s to act as support rails. Pick up a panel with your hands centered on each long edge. Carry it upright, and when you're about a foot from the car, rest the lower edge on your thigh as you slide the top edge onto the 2 × 4 supports. Center the panel on the car roof and use tie-downs to secure the material on the sides and keep it from sliding forward and backward. It's easier to transport more than one panel, because the weight of several stiffens the load.

Installing Paneling

Plywood paneling is relatively easy to install. You can use either nails or adhesives or a combination of both. The large sheets can be difficult for a single person to handle, so seek help. Here are tips to make this project a snap.

Panel Layout Tips

- Lay out the plywood installation so the edges fall over nailers, either studs or backing or nailers installed at the corners.

- Use the chalk box as a plumb bob, or use a pencil and carpenter's level, to make a plumb mark at each panel joint.

- Use a 2-inch-wide paintbrush and a stain that matches the paneling to paint a strip on the wallboard at each corner and joint location. This way, if the paneling shrinks slightly and the joint opens, you will not have white wallboard shining through the crack.

Adhesives

When using adhesives for installation, pad a 2 × 4 wooden block with scrap carpet. As you install each panel, place the 2 × 4 block on the face of the panel, and tap the block with a hammer, moving and tapping over the entire panel surface. This will ensure good contact between the paneling and wallboard base. Some adhesives require that you push the panel in place, against the

nstallation Tips
Plywood Paneling

- Modern paneling is relatively stable and free from expansion and contraction, but paneling that has been stored in an unheated space may have absorbed some moisture. Buy the paneling a few days before beginning installation, and lay the panels flat on the floor. Let the panels remain flat for several days before installation, to allow them to acclimate—that is, come up to room temperature and humidity—before you install the paneling. This will minimize panel shrinkage.

- To secure paneling to a wallboard base, use nails that are color-coded to the paneling, or use panel adhesives. Adhesive installation is best, because you eliminate the nail holes. Use only a few nails at the panel edges to hold the panel in place until the adhesive sets.

- Always position the panels so that finished or factory edges meet at joints. If a cut edge is exposed, so that the light-colored wood subplies are visible, run a carpenter's pencil along the cut to darken the edge. This way, the color of the cut edge will match the color of the finished edges.

adhesive, then pull the panel away to let air reach the adhesive. Then push the panel back against the wall and tack it in place with a few nails to hold the panel until the adhesive sets. Follow the instructions supplied with the adhesive.

Fitting Irregular Panels

Corners at walls or at wall and ceiling can be concealed with wood molding. If you prefer not to use corner molding, position the panel to be cut about 1 inch away from the already installed panel on the opposite side of the corner. Plumb the panel and tack it in place. Now spread the legs of a carpenter's compass enough so the two tips span the distance between the panel to be joined and the proposed cut line. Place the metal-tipped compass leg against the installed panel, and the pencil leg on the proposed cut line on the panel to be installed. Keeping the legs of the compass level, pull the compass down the length of the panel. This will leave a scribe mark on the panel. Cut carefully along this scribe mark and you will have a cut edge that fits tightly against the installed panel's edge.

Laminates/Corian

Laminates

Plastic laminates, popularly referred to by the trade name of Formica, are used as a surface finish on many products, including furniture and countertops for bathroom vanities and kitchen cabinets. Because of the skill level necessary to make countertops, most homeowners buy their countertops prefabricated from cabinet shops. Home centers also offer prefabricated laminated countertops in their kitchen cabinet departments.

Plastic laminates can be installed by the handyman, and are sometimes installed on the walls between countertops and wall cabinets.

Cutting Laminates

Laminates can be cut using a hand-, saber-, table, portable circular, scroll, band, or radial arm saw. To minimize edge chipping when cutting laminates with a saber or portable circular saw, place the laminate with the finish or decorative side down. When cutting laminates with a hand-, table, scroll, band, or radial arm saw, place the laminate with the finish side up.

Edges of laminates can be smoothed with a hand file or sanding block. Trim overhanging edges with a router fitted with a straight or flush laminate-trimmer bit.

To cut the ends of laminates to make an end or butt joint, overlap the ends of the two laminate strips by about 1 inch, then clamp them together with wood strips. Position the two wood strips so they will act as a guide on both sides of the router base. Then use the router to make the double cut, and the two ends will fit tightly together to make a perfect joint.

Shaping Laminates

Laminates are quite brittle, so you must heat them to fit them around a curve. Use an electric heat gun, a heat lamp, or a blow-dryer to heat the laminate and make it easier to shape. Test the laminate for pliability, and do not overheat. Use a laminate roller sold at large home centers to roll and shape the laminate around any curves.

Laminates can be installed over plywood or over bare or painted wallboard or plaster. Your dealer can supply the contact adhesive and notched trowel used for laminate installation. This contact adhesive is applied to both the laminate and the underlayment, usually a plywood or particleboard countertop.

One tricky part of laminate installation is that the laminate will stick on contact with the adhesive, meaning that you cannot move it to adjust the fit. For countertops, apply laminate to the edges first and trim them with a router laminate bit. Use a file to do final smoothing. Apply adhesive to the plywood top, then let the adhesive set so it is not sticky to the touch. Position small dowels or the slats from a venetian blind at 6-inch intervals perpendicular to and along the length of the countertop. Cut the laminate so it overlaps the countertop on all sides. Apply adhesive to the laminate, then position the laminate above the dowels or slats and align it so there is overlap on all sides. Carefully remove the dowels or slats, one at a time, and press the laminate into contact with the adhesive on the plywood. Continue until all dowels or slats are removed and the laminate is in total contact with the countertop. Roll the entire laminated area, trim the edges with a laminate-trimmer bit, and smooth the edges with a file.

The roller on a laminate-trimmer bit may be jammed by laminate adhesive, causing the bit to be pulled into the work piece and damaging it. To avoid this, keep a can of WD-40 handy, and spray the roller of the bit. WD-40 will prevent the adhesive from sticking to and jamming the bit roller.

To clean adhesive from laminate, use a scrap of laminate for a scraper. Then use a cloth dampened—not soaked—with lacquer thinner to remove adhesive residue.

Corian

Corian is the trade name of a plastic coating used on countertops and furniture. Unlike other plastic laminates, which have a colored finish surface and black backing, Corian's colors are continuous through the product, so there is no black corner when the Corian is trimmed for fit where the edges meet.

Corian is more rigid and harder to work than plastic laminates, so it is not a do-it-yourself product. Have an experienced professional install Corian.

Wallcovering

Wallcovering was once known as wallpaper, because that is what the material was made of. Today, the most popular wallcoverings are vinyl, or paper or fabric with a strippable vinyl backing. Vinyl wallcoverings are available prepasted, so you must soak the product in a water trough to activate the adhesive, rather than brush on old-style wheat paste or wallpaper adhesive. Due to the new easy use and because the industry has waged an extensive public education campaign, wallcoverings are now sold for homeowner installation.

To help you learn to apply wallcoverings, decorating centers often hold educational seminars. They will sell you the necessary tools, often in a handy kit, and teach you how to use them.

Once you have learned the basics of wallcovering application, practice by papering a small room that has few obstacles, such as a small bedroom. Because bathrooms may have many obstacles, such as pipes, showerheads, and many corners, avoid bathrooms until you have some experience.

Choosing Wallcoverings

Wallcoverings are available in four basic patterns or designs. These designs are geometric (such as plaids), large prints (bold florals), small prints (small florals), and overall prints (repeated abstract designs with no distinguishable background). Wallcoverings are offered with coordinated fabrics and borders.

nstallation Tips
Wallcovering

Measure the height of the wall from baseboard to ceiling. Be sure there is a full pattern at the top of each wallcovering strip at the corner of the ceiling. Note the drop length, or the distance between matching patterns. Allow for this drop when you cut for length, then cut the strip so it overlaps the ceiling or top molding and the base trim by about 2 inches at each end. This overlap ensures that strips will be long enough to span the entire wall height. Using the broadknife as a guide, cut off the overlaps with the razor knife.

If you are hanging fabric coverings, check the manufacturer's directions. Some materials call for the adhesive to be rolled on the walls, others on the back of the coverings.

If you are hanging prepasted covering, fill the water tray with lukewarm water and place the strips in the water. Let them soak for about 1 minute, then remove and let excess water drain away. Now reroll the coverings with the pattern side in and let the strips stand on a plastic drop cloth for about 10 minutes.

Before hanging the strip, measure out from the first corner the width of the covering, then subtract ½ inch. Mark this measurement on the wall and make a plumb line from base trim to ceiling. To carry the covering to the wall, unroll the strip and fold both ends inward until the ends meet in the middle. This is called booking the covering. When you hang the strip, match one edge to the plumb line, and fold the ½-inch surplus around the corner of the adjoining wall. Use a smoothing brush or plastic smoothing tool to smooth the covering. Use the razor knife to cut off the overlap at the ceiling and base trim. Let each strip set on the wall for about a half hour, then use the seam roller to roll the seams. Use the sponge and clear water to rinse excess adhesive from the face of the covering.

Continue to hang the strips, being sure to match the pattern from strip to strip, until you reach the next corner.

To turn a corner, measure the distance from the edge of the previous strip to the corner, and add ½ inch. The extra ½ inch will be folded around the inside corner. Measure out from the corner the full width of the strip and establish a plumb line there. The overlap at the corners will ensure that the corner seam does not open and reveal bare wall underneath.

Types of wallcoverings include vinyl, foils, flocks, grass cloth, fabric, and embossed.

When you have chosen a wallcovering that you like, ask for a sample book or buy a roll and take it home. Cut a large sample of the wallcovering and stick it to the wall with adhesive tape. Check the effect over a 24-hour period, in varying light situations, to be sure it is exactly what you want.

Wallcoverings are available in single or double rolls. The area covered by a roll will be listed on the package, but waste may reduce the stated coverage by 15 percent, so remember this when buying the wallcovering. Try to estimate your wallcovering needs accurately, buying an extra roll if necessary, to be sure you have enough material.

Use the Right Adhesive

If you are hanging unpasted paper, be aware that different adhesives are needed for different coverings. For example, use Heavy Duty Vinyl Wallcovering Adhesive for hanging vinyl or vinyl-backed coverings. Use Vinyl Over Vinyl adhesive for joints where vinyls overlap, or to apply vinyl borders over vinyl wallcoverings. Use clear adhesive or ordinary wheat paste for flocked or fabric coverings. Ask your dealer to help you choose the right adhesive for your job.

Tools Needed for Wallcovering

Tools needed for wallcovering installation include a paint pail, large sponge, water tray, bubble stick (combination yardstick and level), paint tray and roller, smoothing brushes, plastic smoothing tool, wallcovering scissors, razor knife, seam roller, and 6-inch broadknife or putty knife.

You will also need a flat work surface

RULE OF THUMB

Maintaining a Sharp Cutting Tip

To cut wallcovering without snagging or tearing it, keep the knife blade sharp. Buy a razor knife with multiple breakaway tips, and break off the old cutting tip frequently to ensure that the knife is always sharp.

on which to place the wallcovering strips. Rent a wallcovering table or place a sheet of plywood over two sawhorses to make a flat work surface.

Removing Wallcoverings

Most modern wallcoverings are strippable, meaning that you can simply lift one corner of the covering and pull the entire strip from the wall. Wash the wall carefully to remove the old adhesive.

If the old wallcovering is not strippable, use warm water mixed with a covering removal product, available in paint departments, to soften the old adhesive. If the old covering is vinyl, use a tool called a Paper Tiger to cut through the vinyl. Apply the water/remover mix with a sponge or sponge mop. Use a paint scraper to test the covering. If the adhesive has softened, the covering will come away easily. If the adhesive has not softened, do not chisel away at the covering. Simply apply more water/remover mix and wait until the adhesive has softened. The covering will peel away in full strips.

8 Ceilings and Floors

If you measure the total interior surface of your home you will find that about half of the interior square footage is in the ceilings and floors, with the walls making up the other half. When decorating, keep in mind that the walls are broken up by windows and doors, making it easier to disguise and conceal wall problems. But ceilings and floors are the largest unbroken surfaces in the house, so money spent on decorating these surfaces has a magnified impact on the appearance of the home's interior. It is important to remember that newly decorated ceilings can produce a startling transformation of the home's interior, and that problem floors can be covered up quickly and inexpensively with carpets or other flooring.

Ceilings

Plaster Ceilings

In houses built before 1950, plaster was the most common ceiling finish. As the wallboard industry matured, wallboard replaced plaster in all but the most expensive houses.

Plaster was applied over a base of lath. In houses built before 1950, the common plaster base was wood lath. Wood lath consists of small strips of wood, about $3/8$ inch thick, spaced about $3/8$ inch apart. These gaps between the wood lath are called the key strips. When troweled over the wood lath, the soft plaster was forced between the wood strips and locked to them via these "keys." The plaster did not stick to the wood lath itself, but was secured in place by the keys.

In time, plaster keys deteriorate and the plaster surface fails. If the plaster is loose over small areas, you can use large-headed plaster screws to secure the plaster to the lath or framing. When the plaster keys fail over large portions of the surface, the plaster must be removed and the surface replastered.

By the 1950s, most wood lath was replaced by rock lath. Rock lath is similar to wallboard, but is $3/8$ inch thick, 16 inches wide, and 48 inches long. Plaster will bond directly to rock lath, so plaster failure over a wide area is rare in houses with rock lath. Only a secondary problem, such as water damage from a leak in the roof or plumbing, will cause extensive failure of the plaster surfaces.

The third type of plaster lath is a steel mesh known as wire lath. Wire lath is not commonly used for residential plastering, but is found in houses of the mansion class.

Troubleshooting Plaster Ceilings

Most plaster problems can be solved by patching. Before undertaking plaster repair, press with the palm of the hand against the plaster. Do this wherever there are water stains, and on both sides of any cracks in the plaster. If the plaster feels loose or spongy to the touch, the plaster has loosened over the area, and the loose plaster must be removed before repairs can be done. If the plaster appears to be tightly bonded to the lath, proceed with repairs.

Patching holes in plaster. Patch plaster is nonadhesive, and you must cut a piece of wire lath to fill the hole and provide the plaster with a base to which it can key. Buy the wire lath at masonry supply stores. Secure the wire lath in place with staples or nails. A good substitute for patch plaster is to use quick-set wallboard compound, which is adhesive and sticks to the old lath and plaster better than patch plaster. Several coats of either patch plaster or wallboard compound may be needed to fill the patch to level.

Patching plaster cracks. When patching plaster cracks, assume that all cracks are structural. Structural cracks are those that will reopen the next time the framing moves, that is, when the temperature or humidity changes. Use paper wallboard tape—not fiberglass tape—to reinforce all plaster cracks, and finish the taped cracks as you would finish the joints in wallboard.

New Skin for Plaster Ceilings

Removing and replacing old ceiling plaster is a dirty and difficult job. If there is blown-in or loose-fill insulation above the ceiling, the insulation also must be removed. But there are ways to resurface the ceiling without disturbing the insulation and creating all the dust and mess. If plaster ceilings seem to have reached the end of their life, there are several options to removal and replastering.

If old plaster is badly cracked or crumbling but the ceiling is still smooth and level, you can apply a layer of wallboard directly over the old plaster. Use 1/2-inch-thick wallboard to cover the plaster, and use wallboard screws to secure the new wallboard in place. Be sure the screws are long enough to penetrate through the wallboard, the 3/4-inch layer of lath and plaster, and into the ceiling joists. Wallboard screws that are 2 inches long will do the job.

If plaster is buckled and wavy, apply 1 × 2 furring strips at 16-inch intervals, and use wallboard screws to secure them to the ceiling joists. Use wood shims to level the furring strips, and screw the wallboard to the furring strips.

A third option for older houses with high ceilings is to lower the ceilings. To bring ceilings down to modern heights, install new ceiling joists at a level point 8 feet above the floor. Apply new wallboard over the new joists. Or, install a suspended (tile) ceiling below the existing plaster ceiling (see "Suspended Ceilings," on page 37).

Wallboard Ceilings

The most common problems with wallboard ceilings include popped or loose nails or screws, and water damage from roof or plumbing leaks. If the wallboard has sagged or appears soft from water damage, you must remove and replace the damaged wallboard. If wallboard ceiling panels appear loose due to popped nails, renail or screw the old wallboard to pull it tight to the ceiling joists, then redrive the old nails. If ceilings are finished with smooth paint, you can simply apply three coats of wallboard cement over the screw heads (allow to dry

RULE OF THUMB

Wallboard Ceilings

Because wallboard ceilings now are supported by roof trusses set 24 inches on center, and because new ceiling insulation requirements pose a greater weight problem, use stronger, stiffer $\frac{5}{8}$-inch wallboard rather than $\frac{1}{2}$-inch wallboard for new ceilings.

between each coat) and repaint the ceiling. If the ceilings are textured, you may prefer to cover the old wallboard with a new layer of wallboard.

If wallboard ceilings are sagging between the ceiling joists, apply furring strips at 16-inch intervals over the ceiling. Use wood shims to shim the furring strips level. Then apply a layer of $\frac{1}{2}$-inch wallboard over the furring strips and finish the wallboard.

Laminated Wallboard

Certain desirable characteristics of a ceiling depend on the mass or thickness of the ceiling finish material. Thus plaster, which is $\frac{3}{4}$ inch thick, is both more soundproof and fireproof than $\frac{1}{2}$-inch wallboard.

If you desire better soundproofing or fireproofing from a ceiling, you can achieve this by using multiple layers of wallboard to meet any requirement you choose. For example, rather than use one layer of $\frac{1}{2}$- or $\frac{5}{8}$-inch wallboard, use two layers of $\frac{3}{8}$-inch wallboard to achieve the same mass ($2 \times \frac{3}{8}$ inch = $\frac{3}{4}$ inch) as plaster. This will be both more soundproof and fireproof than single-layer wallboard. True laminated

wallboard systems consist of one layer of $\frac{3}{8}$-inch wallboard nailed or screwed to the framing, and a second layer of $\frac{3}{8}$-inch wallboard secured to the first layer of wallboard using wallboard adhesives. Temporary nailing is used to hold the wallboard in place until the adhesives set. See your wallboard dealer for complete instructions on laminated wallboard procedures.

Suspended Ceilings

Suspended ceilings are often installed in basement recreation or laundry rooms, where below-joist obstacles, such as plumbing pipes, wiring, and heating ducts, may make installation of wallboard more difficult. Suspended ceilings are also popular in kitchens, where a suspended ceiling is installed at soffit height, just above the top edge of the wall cabinets. Often, an inexpensive fluorescent light is installed against the ceiling joists, and a clear plastic lighting panel is inserted in the metal ceiling grid below the fluorescent bulbs to provide indirect lighting.

If you decide upon a suspended ceiling, the material will be supplied along with manufacturer's instructions for installing

RULE OF THUMB

Using Multiple Layers of Wallboard

When using wallboard to patch a hole in plaster, you will find that $\frac{1}{2}$-inch wallboard does not fill the hole to make a level patch. Instead, use one or more layers of $\frac{3}{8}$-inch wallboard to fill holes in plaster.

your particular system. Following is a brief description of the steps you will follow to install a suspended ceiling (see figure 8.1).

● To begin installation of a suspended ceiling, you must first locate the ceiling joists concealed by plaster or wallboard. Use an electronic stud finder to locate the concealed joists. Find both ends of the joists, then use a chalk line to mark the center of the joists along their entire length.

● For most grid systems, the long sides of the 2 × 4 ceiling panels run parallel to the shorter walls of the room. Lay out the plan so the main runners are perpendicular (at right angles) to the joists. For proper balance, the border or cut panels should be the same size. To lay out the plan, divide the room length by the panel length. For example, if the room is 15 feet long and 12 feet wide, divide the length (15 feet) by the panel length (4 feet) to get 3 with a remainder of 3. Add this 3-foot remainder to the length of a 4-foot panel to get 7 feet. Divide 7 by 2 to get 3½ feet. The room will have two full-length 4-foot center panels with two border panels that are 3½ feet each. Use the same

Figure 8-1 Layout for a Suspended Ceiling

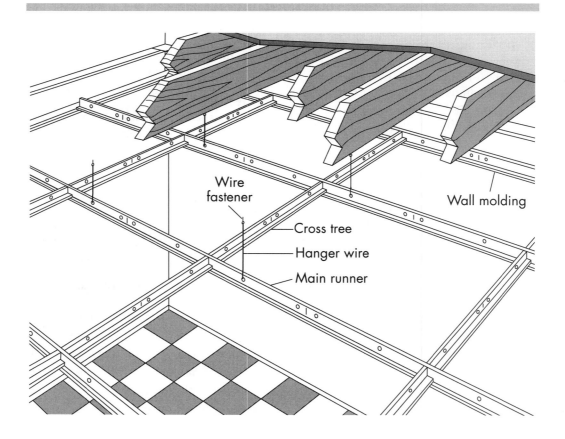

Wire fastener

Wall molding

Cross tree

Hanger wire

Main runner

arithmetic to find the size of the border panels. For example, take the width of the room (12 feet) and divide it by the width of the panel (2 feet). This comes out to be an even number, 4. So your plan will have four rows of tiles. Each row will have two full-size tiles and a 3½-foot tile at each end.

- Next, install the wall molding. Make a mark at the height of the new ceiling. Then add the height of the wall molding. Use a carpenter's level to make a level line around three walls, then use a chalk line between the corners of the fourth wall to mark it. Position the tops of the wall moldings at the lines and nail them in place, driving nails 16 inches apart or at each wall stud location.

- Consult your plan to find the location of the main runners. These usually will be located at 4-foot intervals and run perpendicular to the ceiling joists. Mark the location of the runners on the wall molding. Note that in the above example the main runners closest to the wall will be located 2 feet from the wall to accommodate the border panels. Then stretch a chalk line across the room from one of these marks to the matching mark on the opposite wall and snap a layout line on the ceiling. If the joists are exposed, the chalk line will make a mark on each joist. Use the chalk line to mark the location of the main runners.

- To locate the position of the hanger wires, measure the distance of the border tile you calculated for the plan along the main runner layout line. In the example, the border is 3½ feet long, so you should mark the joist that is closest to this position. Install a wire fastener or screw eye in the center of the joist where the chalk line and joist

intersect. Do the same at the other end of the room and then install the remaining wire fasteners at 4-foot intervals between the two end hangers you've just installed. Repeat these steps to install the other wire fasteners for the other main runners.

- Install a hanger wire on each of the wire fasteners. These wires should be long enough to extend at least 6 inches below the wall moldings. Thread the wire through the eye of the hanger and then twist each wire around itself at least three times.

- Mark the ceiling height on the hanger wires. Stretch a chalk line along the row of hanger wires. Hold the chalk line even with the upper edge of the wall molding. Snap the chalk line to mark the wires, then take pliers and bend the hanger wires at a 90-degree angle ¾ inch above the chalk mark on the wire. Repeat this along each row of hanger wires.

- Trim the first main runner so the slot for the cross tree is the proper distance from the wall. The cross trees fit into slots in the main runner that are on 2-foot centers. In the example the border tile is 3½ feet long so the first cross tree must be 3½ feet from the wall. The main runner slots are on 2-foot centers, so start at the second slot from the runner end and measure 3½ feet from this slot toward the closest end. Mark this spot and cut the end of the runner off at the mark. This will place the cross tree slot 3½ feet from the end of the runner.

- Place the cut end of the main runner against the wall so it rests on the wall molding. Then thread a hanger wire through the closest hole in the upper edge of the runner. You may have to adjust the bend in the wire to level the runner. Bend the hanger wires around the main runner loosely since they will probably have to be adjusted.

- Slip the tab of a full section of main runner into the slot at the end of the first section you've just installed and attach the hanging wires to the new section. Continue adding sections of main runner until you reach the other end of the room and then cut a section of main runner to fit.

- Install the other main runners in exactly the same way and then install the full-size cross trees joining the main runners.

- Cut cross trees to fit between the main runners and the wall molding to complete the grid.

- Check that the grid is level. You may have to adjust the hanger wires. Then make sure all the hanging wires are secured by twisting the wire around itself at least three turns.

- Lay in the ceiling panels.

Ceiling Moldings

Use ceiling moldings to cover the joints between ceilings and walls, or for decorative purposes. Moldings are available in either wood or polystyrene plastic. Use cove moldings or a combination of several molding shapes to gain the desired effect. For directions on installing moldings, see chapter 3.

Floors

In houses with wood floors, construction has changed little through the years. In 1950 plywood began to replace 1 × 6 boards for use as the first layer of boards nailed directly to the floor joists, called sub-floor sheathing. The floor system of a wood-frame house consists of basement or foundation piers or posts to support a center beam set down the middle of the house. Resting upon the post/beam support are floor joists, 2 × 8s, 2 × 10s, or 2 × 12s, depending on the width of the house or the span of the joists. At floor openings for stairways, there are doubled or trimmer joists along the length of the opening, and double header joists across the width of the opening. These doubled-up joists strengthen the area around these openings.

Between the joists there may be short pieces of wood called bridging, used to transfer the load from joist to joist and to stiffen the floor. Bridging also prevents the joists from twisting or warping, and thus holds the joists perpendicular to the beams and support walls beneath the joists. This entire structure comprises the floor framing.

The next floor component is the sheathing or subfloor. In houses built before 1950, the subfloor sheathing is 1 × 6 or 1 × 8 boards. Beginning in the 1950s, plywood was used as floor sheathing. Plywood is both stronger and faster to work with than boards. Over the plywood, either strip hardwood flooring or another layer of plywood or particleboard is laid. The plywood is the underlayment for flooring such as carpeting or vinyl sheet goods. In most cases, a layer of builder's paper, sometimes called red rosin paper, is laid between the subfloor and the hardwood flooring or the plywood underlayment. (Builder's paper is used only if the hardwood flooring is nailed to the joists, not in adhesive applications.) The rosin paper helps quiet the floor by preventing any friction noises when the floor sheathing and finish flooring move against one another. If the finish flooring is hardwood strips (usually either oak or maple), then hundreds of joints are created between these hardwood strips. To finish off the joint between walls and floor, base molding, also called shoe, is nailed around the perimeter of the floor.

The homeowner must understand that each component in this floor structure is a possible source of flooring squeaks. Strip hardwood flooring can flex and rub against adjoining strips when a person walks upon the floor. Or the strip flooring or plywood can move against the subfloor boards or plywood sheathing. The sheathing in turn can move and rub against the floor joists, the floor joists can creak if they sag under a person's weight, or cross bridging can rub against adjoining bridging and cause squeaks.

Before you install new flooring, you should correct squeaking floors and stairs, and other floor problems. This can be done by having one person move over the floor while another person marks the squeaking areas. The loose or squeaking areas can then be renailed. If you plan to sand a hardwood floor, try to secure the flooring to the subfloor from the basement side by driving screws upward through the subfloor and into the finish flooring. To avoid driving the screws completely through the surface of the finish flooring, be sure to use screws of the proper length—that is, screws that are not longer than the combined thickness of the subfloor and flooring.

If you are covering the floors with sheet goods or carpet, you can drive nails or screws downward through the face of the wood flooring, into the floor joists below, to quiet floor squeaks.

Hardwood Flooring

For several decades, it was fashionable to cover beautiful hardwood floors with carpet or vinyl flooring. Today many homeowners have rediscovered the beauty and durability of hardwood floors. For variety and accent in new houses, one or more rooms often have hardwood floors.

Laying hardwood floors once was work reserved only for the professional. Installation included fitting and nailing the floor strips in place, then sanding, filling, staining, sealing, and finishing the floor. Today, rental tools make installing hardwood flooring a project that experienced do-it-yourselfers can do themselves.

Strip flooring is blind-nailed in place over a smooth subfloor. To make this process easier, a special nailer has been developed that fits over the edge of the flooring and drives a nail through the edge of the board at the proper angle (see figure 8.2). The power nailer and the special nails that fit into this tool are available from most rental centers.

Today, hardwood flooring materials are manufactured under the most careful temperature and humidity controls, and the flooring is sanded, sealed, and finished at the factory. The homeowner can install the flooring and eliminate the jobs that are both difficult and messy—the sanding and finishing.

Hardwood flooring is available in a number of styles, including strip flooring in a variety of widths, up to plank dimensions, and in a variety of styles, including parquet. Parquet squares are manufactured in 1-foot-square tiles designed to look like four 6-inch-square tiles.

Shop for the particular style of hardwood flooring you like at your local home center or tile outlet. Be aware that some hardwood floor products are not approved for basement or below-grade installations.

Because the flooring is a manufactured wood product, moisture content is carefully controlled from the factory to the consumer. This means that the flooring contains the same moisture content as wood furniture or prefinished cabinets, that is, about 7 percent. However, it is possible that

Figure 8-2 Installing a Hardwood Floor

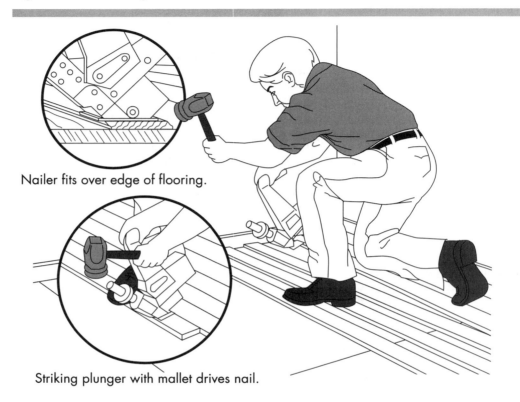

Nailer fits over edge of flooring.

Striking plunger with mallet drives nail.

the prefinished flooring may pick up a bit of moisture in the shipping-handling-storage process. It is recommended that you haul both the adhesive and the hardwood flooring home, open the flooring cartons, and leave the materials in the room where they will be installed. This process, called acclimation, permits the products to reach the room temperature and humidity level before installation. This will help prevent the flooring from either shrinking or expanding after it has been installed.

Hardwood flooring was once nailed in place, but most of the prefinished flooring products aimed at consumer installation today are secured in place with flooring adhesives. To install the flooring, use one of the latex-based adhesives that are low-odor and solvent-free. The adhesives may be applied over a base of concrete, wood, plywood, or particleboard.

The key to a successful hardwood adhesive application is to be sure the base surface is clean and free of wax. Use a wax-stripping product to remove the wax from vinyl floors. Before installing any flooring over existing vinyl floors, be sure the vinyl is tightly adhered to the subfloor. To apply the adhesive, use a notched trowel of the type recommended by the flooring manufacturer.

The starting point in the room will depend on which type of flooring you are laying. The directions for installing your own particular flooring will be included in the product package. For parquet tiles, measure to the center of the room from both directions, and chalk-line the centerlines for both room width and length. As recommended by the manufacturer's directions, apply the adhesive to the base surface. For most applications, you must let the adhesive set for 30 to 60 minutes to "tack" or become slightly set before installing the wood flooring.

To allow for any slight expansion of the wood floor, leave a ½-inch gap between the flooring and the wall on all four sides. Use base or shoe molding to cover the crack between the flooring and the walls.

Ceramic Tile Flooring

Ceramic tiles for floors are most often used in bathrooms, kitchens, or entry halls. Shop for ceramic tile at specialty shops, where you will find a wide variety of choices for size, style, and color. Some ceramic tiles are intended for wall application only, so be sure to specify that you are shopping for floor tile when talking with your dealer.

Ceramic tile floors once were set only on a 1-inch-thick bed of concrete. Today the tiles are often set over prepared ply-

wood floors, bonded with a concrete product called Thinset. If you decide to lay a ceramic tile floor with Thinset, be sure that the plywood base is stiff enough so that it will not flex under a person's weight. A floor that flexes under foot traffic will cause the ceramic tile to pop loose from the plywood base.

After the tile adhesive has set, you must apply grout to fill the cracks between the tiles. Use a rubber squeegee to spread the grout over the floor, then use a wet sponge to wipe away excess grout.

Laying ceramic tile floors requires use of specialty tile tools. If you decide to do the job yourself, you can rent the needed tools from the tile store.

Vinyl Floors

A common concern in the flooring industry has been the danger of asbestos exposure from removing old flooring. According to the Environmental Protection Agency (EPA), the danger from asbestos is in breathing the airborne asbestos fibers. For this reason, at replacement time it is best to simply leave any flooring material in place rather than remove it. Manufacturers are now banned from using asbestos in new home products, so if you cover over the old product, there will be no future danger from asbestos exposure. Following are several ways to install new flooring without removing the old product.

Installing New Flooring over Old Flooring

If the old flooring is in bad shape, install a plywood underlayment for new flooring to cover up the old flooring. This is the best approach for many reasons. For example, if you remove the old flooring, the old black adhesive ("cutback") may bleed through new vinyl flooring. The best technique is to

RULE OF THUMB

Good Adhesion

After hardwood floor installation, wait 24 hours for the adhesive to set. Then roll the floor in both directions with a heavy roller (75 to 100 pounds). Floor rollers can be rented at tool rental outlets. After rolling, wait another 24 hours before moving furniture or appliances onto the hardwood floor.

install a new layer of ⅜-inch luaun mahogany plywood over the old flooring. Just nail the plywood in place following the directions of the flooring manufacturer. Then use a floor filler such as Dependable to fill and smooth the joints in the plywood. Dependable is a product you will not find at home centers, so check in the Yellow Pages for a dealer who sells materials to professional flooring installers to find Dependable or a similar floor filler.

If built-in cabinets or appliances make a ½-inch rise (subflooring and flooring) in the floor level a problem, you can cover directly over the old flooring. Use a wax remover to remove any floor wax, then trowel two coats of Dependable over the entire floor. If the flooring is embossed, the pattern may show or "photo" through the new flooring. Ask the professionals' dealer to supply you with an embossing compound. Apply two or more trowel coats of embossing compound to smooth the surface. Then install the new flooring.

Inlaid versus Vinyl

Inlaid products, which have several layers of vinyl to produce an embossed pattern,

compare in quality to the old linoleum of yesteryear. The product is heavy and, because of its weight, is usually available only in 6-foot-wide rolls. This means that you will have edge seams if you have inlaid product installed. (Hire a professional.) The plus is that the heavy weight of the inlaid product translates into durability. Also, the pattern and color are continuous through the entire thickness of the inlaid product, so any wear is less noticeable. If you choose an inlaid product, be sure the installer uses the proper adhesive for the job.

Inlaid flooring products are more expensive, for both the material and labor. A good example of this superior product might be Armstrong's Designer Solarian. Expect to pay about $30 per square yard for inlaid material, plus around $8 per yard for labor and floor prep (cost varies by region).

Many new vinyl flooring products now approach the quality of inlaid products. Vinyls with thinner wear layers, often called builder grades, may look good when new but will not provide long wear. Use lighter-weight products only in low-traffic areas.

Vinyl flooring offers a wide range of colors, patterns, and designs. Because vinyls are made in 12-foot widths, they can be seamless when laid in rooms less than 12 feet wide. There still may be seams at doors or on wider rooms (called "fill"). Seams that are not sealed let air or water enter, damaging the adhesive so the edge of the vinyl will curl back and ruin the appearance of the floor. Be sure the installer uses a seam sealer to join the material together at the seams.

RULE OF THUMB

Buying Quality Sheet Flooring

Vinyl floor products have surface layers between 10 and 30 mils thickness. Buy a vinyl product with a wear-layer thickness of at least 15 to 20 mils.

Vinyl Tiles

In the past, vinyl tiles were called Vat or vinyl asbestos tile. But in 1980 asbestos was banned from home products, so today they are called Vct, for vinyl composition tile.

The drawback to using floor tile is that there will be 4 feet of seam per 12-by-12-inch tile. Water from floor washing can seep into the tile seams and break down the tile adhesive. This is especially common around sink areas in the kitchen or bathroom.

On the plus side, vinyl floor tiles are easy to install, even for the homeowner. The tiles are durable and provide flexibility of design. For example, you can combine two colors, as in a black-and-white checkerboard pattern. Creative freedom in borders is also possible with tile. Another plus for vinyl floor tile includes ease of repair; damaged tiles can be removed and new tiles put down in their place. Vinyl tiles are available as "dryback" that must be applied with adhesives, or in peel-and-stick tiles that have the adhesive on the back of the tiles. If you cover a floor with vinyl tile, save a few leftover tiles for future repairs.

Vinyl Floor Maintenance

On vinyl tile or inlaid flooring, use a good quality floor wax to protect the flooring and keep the floor looking its best. Use a water-and-vinegar solution to clean dirty floors. Do not mop liquid onto the flooring; damp-mop only to avoid water penetration into the seams. Apply a good floor wax per the manufacturer's instructions. Most waxes will yellow with age, so use a wax stripper to remove old wax and recoat with the floor wax of your choice. Floor wax is especially needed on vinyl tile floors, because the wax helps prevent water from entering the seams and attacking the tile adhesive.

"No-wax" vinyl floor products were designed to retain their sheen while eliminating the chore of applying floor wax. On no-wax flooring, just damp-mop with a solution of warm water and vinegar. Some homemakers prefer to use a 'mop and glow' floor product to enhance the gloss or sheen of the floor finish. Be aware that if you apply such products, you are applying a thin coat of floor wax, and you will have to use a wax stripper when the product turns yellow with age.

Area rugs are often used by sinks or kitchen doors. Never lay a padded area rug over a vinyl floor. The padding on the rug may contain petroleum products that are incompatible with the vinyl flooring, and may leave a permanent stain in the vinyl. If direct sunlight falls on the rug, the problem is compounded by the heat.

Carpet

Carpeting and other flooring may be the single most important factor in your decorating scheme, because flooring offers a unifying point that ties together all the color and texture in a room. Carpet offers color, texture, and floor insulation, and provides underfoot comfort and easy maintenance. Because carpets are available in such wide color and style options, it may be difficult to reach a buying decision. To ensure that you make the right decision, take time to learn the basics of carpet construction and care, and choose your carpet supplier carefully.

Installing carpet is not too difficult for the average do-it-yourselfer to handle. There are specialized tools you can rent at most rental centers. The most important factor to consider when deciding whether you should tackle a carpet installation is the cost of the carpet.

Inexpensive indoor/outdoor carpet is easy to install, and since it is rather inexpensive, the cost of professional installation can represent the major portion of the bill. However, top-of-the-line carpet goods are expensive and difficult to work with. In this case, the cost of professional installation is a much smaller portion of the total cost. Think about this when planning.

Choosing a Carpet Dealer

Finding a reputable carpet outlet should not be difficult. Most national department stores have well-earned reputations for honesty, meaning you will get what you pay for and may even get a bargain if you shop the sales. Local carpet dealers who have served their neighborhoods for years also have earned a place as honest businesses. Be wary of the bargain carpet outlets that profess to undersell their competition. A deal that sounds too good to be true is probably neither good nor true.

Ask the salesperson to show you the carpet specification books for any carpet you are interested in. If the dealer cannot supply specification books, the products sold may be carpet "seconds" or carpets with manufacturing defects. Some carpet defects may be only in the yarn or pattern, but some defects involve problems with adhesives or other materials that could affect the durability of the carpet. If you buy seconds from a discount dealer, you may soon be vacuuming up your carpet investment.

In recent years there has been a rapid growth in the number of 800-number carpet ads for companies that sell carpets by phone. In some cases, the carpet dealer will send you carpet samples from which to make your selection. Other dealers ask you to do the legwork in your own community, checking local suppliers' samples so that you can provide the manufacturer's name and the style and color of the carpet when you call.

Most of these dealers will offer carpets at or near wholesale prices. You will also save the sales tax if you are buying carpet from out of state, but you will have to pay the shipping charges. Many of these carpet-by-phone dealers are in the area around Dalton, Georgia, where the process of tufting was begun and developed to its current state of the art.

When you have decided upon a dealer and have obtained carpet sample books, study the carpet specifications. Note especially the brand name, warranty information, style, density, weight, and pile content of the carpet. These are the factors that can affect carpet quality.

Carpet Style

Although woven carpets are still available, most carpets today are made by the tufting process. In this process, tufts of carpet yarn or fiber pierce through the backing material and form the carpet pile. If the pile loops are uniform in size and height, the carpet is called a level-loop pile. Berber styles, for example, are level-loop carpets that have tight loops and maximum fiber density. Carpets with loops of uneven height are called multilevel loops. If a carpet has cut loops and even height, the carpet is called a cut loop, but if loops are uneven in height and some are cut, the carpet is called a random shear. When all loops are cut to a uniform ½-inch height, the carpet is called a saxony, while the same carpet with a longer pile or loop may be called a velvet, plush, or textured plush.

Carpets that have straight tufts blended with twisted or curled tufts are called friezes. These friezes or twists, with their curled tufts, offer a resilient textured pile that resists matting and does not show footprints. These carpets are also called trackless. Check to be sure the twisted tufts are heat-set to preserve tuft resiliency.

The most durable carpets are the various loop piles, and they are the best choice for high-traffic areas such as halls, stairways, or family rooms.

Density and Weight

Density refers to how tightly packed the carpet fibers are, or the number of carpet fibers per square inch. This will be shown in the carpet specifications book. Density is

Choosing Carpet Weight

Check the carpet's pile yarn weight. For saxonies, berbers, or plushes, select a carpet that does not have less than a 40-ounce pile weight, as more pile weight is better. Buy commercial carpet with at least a 26-ounce pile weight. Note that this weight is the weight of the carpet pile only; there will also be listed a total weight of the carpet, which includes not only the weight of the pile but also the weight of the carpet backing.

shown as "stitches/inch" (spi) or "stitches/3 inches." For residential carpets, expect a density of 9 to 10 spi, 5 to 6 spi for berbers. For level-loop or commercial carpets, expect 10 to 12 spi. After counting the stitches per inch, bend the carpet at a 90-degree angle, as the carpet would be bent around the nose of a stair tread. Look to see how much carpet backing is "peeking" at you. The less visible the backing, the more stitches per inch (the higher the pile density) and thus the more durable the carpet.

Pile Content

Also listed in the specifications book will be the type of fiber used in the carpet. The listing may be shown as "pile content," "pile yarn," or simply "pile." This is also an important gauge of carpet durability.

Carpet yarns include nylon, olefin, polyester, polypropylene, and natural fibers such as wool or cotton. Wool carpets have become rare, because they are more expensive than the man-made fibers. Some early polyester pile carpets were prone to matting and early wear, but dealers now say that today's polyesters are durable and more crush-resistant. Often, polyester fibers are combined in a fiber blend, such as a nylon-polyester carpet.

Olefin may be the most durable of the fibers, but pure olefin is coarse and rough to the touch, so it is often combined with nylon. Tough berber carpets are available in nylon-olefin blends or in pure nylon. Nylon pile combines such features as softness, ease of cleaning, resistance to matting, and good durability, so 100 percent nylon fibers are always a good choice.

Carpet Pads

Because the pad cushions the carpet from premature wear and provides underfoot comfort, the pad choice is as important as choosing the carpet. A carpet pad that is too thick may interfere with good personal balance and may be a tripping hazard to the elderly or infirm. A pad that is too thin will cause premature wear of the carpet and may even cause the carpet seams to split apart. The best way to check a carpet pad is to lay samples of both pad and carpet on the floor, as they would be when laid, and walk or stand on them to test underfoot comfort.

Pad materials include felt, prime (single-color urethane foam), rebond (multicolor urethane foam), sponge-rubber and foam-rubber pads. For both economy and performance, choose a 6- to 8-pound rebond pad for most of your carpet installations. (Special situations, such as steps, may call for a thinner, denser pad. Trust your installer's advice.) Rebond pads are made from multicolored foam particles, bonded together with adhesive. Rebond pads will not "bottom out," or lose their resiliency, as inferior pads may do.

Selecting the Carpet

When you have read the specifications for the carpet samples, it is time to make your final selection. The first priority is to consider your own lifestyle. A busy family will require carpets that are durable and easy to clean, while a working couple or retirees may prefer a more luxurious carpet, with durability a secondary consideration. While halls and stairs require a durable carpet, bathroom carpets must also resist stains and be easy to clean. Ask your dealer if you need help to make the final carpet selection.

Carpet color is yet another consideration. Bring home a large carpet remnant (small samples do not show well) and check the carpet in varying lights from morning to night. Color shifts will occur between sunlight and artificial lighting, so colors can be deceiving.

When considering carpet price, compare only carpets of the same style and construction. Whatever your choice, remember that proper cleaning and care can greatly extend the life of any carpet. The following tips may help you to get all the service your carpet was made to deliver.

Carpet Care and Cleaning

Carpeting is a major decorating investment, so follow good procedures in caring for and cleaning your carpet. Neglect can ruin even a quality carpet, while consideration and routine cleaning can extend the life of even an inexpensive product. Here are tips to help you extend the life of your carpet.

- Buy the best-quality carpet you can afford, one with stain protection built into the product.

- Pad quality is important. Beware of dealers who sell you a good pad and install a cheaper product.

- Use commercial-type floor mats at entry doors, and remove your street shoes when you come inside, to keep dirt outside.

- Don't walk barefoot on carpet. Natural skin oils will stain a carpet and cause it to retain dirt.

- Ban food and drink from carpeted areas, and clean up any spills immediately.

- Install a high-capacity central vac or buy a quality upright vacuum cleaner, one with beater bars to beat the dust from the carpet fibers. Cheap vacuums can't lift out the grit that will cut the carpet fibers.

- Vacuum often. High-traffic carpets should be vacuumed daily.

- Direct sunlight can cause premature fading. Close drapes to shade the carpet.

- To steam out any dents from furniture legs and the like, hold a steam iron about 2 inches above the carpet. *Don't place a hot iron directly on nylon or other plastic fibers.*

- Have carpets professionally cleaned. Small rental carpet-cleaning units leave detergent and dirt residue in the carpet fibers, causing the carpets to soil quickly after cleaning.

- The built-in protective barrier in your carpet will break down with use and cleaning, and you should renew carpet protection to keep the carpet clean. Each time the carpet is cleaned, have it sprayed with a protective barrier such as Scotchgard.

- Save any manufacturer's care or warranty information.

- Save a good-size carpet remnant for future repairs.
- There are several good all-purpose spot cleaners available. Ask your dealer for a product such as TECH or Spot Shot.

Indoor/Outdoor Carpet

Indoor/outdoor carpet is a highly durable carpet that can be used either indoors or outdoors. This all-plastic product does not fade or deteriorate when exposed to the harshest weather. The carpet can be used outdoors to cover a patio, stair landing, or porch. It is an excellent product for covering the floor of a boat.

When using indoor/outdoor carpet to cover a porch or steps, ask your dealer to help you select an all-purpose latex carpet adhesive. This adhesive will hold even in the worst summer heat or winter cold. The carpet adhesive can be applied using an inexpensive notched mastic trowel, available from your carpet dealer.

9 Doors and Windows

Windows provide ventilation and sunlight, and when used as a remodeling tool, modern windows can have a dramatic effect on the architectural statement of a house. For example, consider the standard story-and-a-half tract home. Replace older double-hung windows with modern multiple-light windows, and the fifties picture window with a large bow or bay window, and you have a Cape Cod bungalow. Aside from style, consider the impact modern vinyl or aluminum-clad windows can have on both the energy conservation and the reduced maintenance of a house.

Doors, too, can have a dramatic impact on a house. One remodeler credits both his quick sale and a handsome profit to the fact that he installed a hand-carved oak entry door on a house he remodeled for sale. The improved curb appeal provided by a striking entry door is difficult to measure.

Because doors take up less square footage, the energy savings for upgrading doors is not as great as for windows. Still, insulated foam-core steel doors offer better energy efficiency than a standard door/storm door combination. The steel doors can be finished in a wood grain to resemble wood while providing security and energy savings.

Doors

Parts of a Paneled Door

The two vertical pieces on the right and left of a paneled door (see figure 9.1) are called stiles; the top and bottom horizontal pieces are called the top and bottom rails. The horizontal center rail is called the lock rail, and any additional rails are called intermediate rails. The center pieces are called panels or raised panels, depending on their configuration. Vertical pieces placed between the side stiles are called mullions. Some doors have glass panels called lights.

Parts of a Door Frame

The parts of a door frame include the vertical side pieces, called side jambs; the top piece, called the header jamb; and the bottom pieces, called the doorsill and the threshold. The doorsill and/or threshold may be wood, usually durable oak, or aluminum.

The jambs of an exterior door are usually rabbeted or grooved to form doorstops. The rabbeted stop helps prevent air infiltration around exterior

Figure 9-1 Parts of a Paneled Door

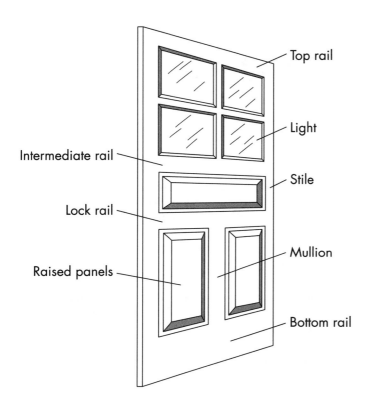

Exterior Doors

Both exterior and interior doors are shipped prehung, meaning the door is hung in the frame with all hardware except the lock in place, and with weather stripping installed. The trim or casing is installed on one side of the door, with exterior trim added after the door is installed in the rough opening.

Although larger custom doors are available, standard entry doors are 1¾ inch thick and 80 inches high. Although codes permit using 32-inch-wide doors (2-8), for easy entry and for access when moving furniture or appliances, main entry doors are usually 36 inches wide (3-0). The minimum width for other exterior doors is 30 inches (2-6), but these narrow doors do not provide access for the handicapped, and will prove frustrating when you must try to carry packages or move other objects through them

Depending on the climate, the entry door may be a wood door with a screen door or combination storm/screen door on the outside. If the house is air-conditioned during warm weather, and the door is not left open for ventilation, carved wood doors with no screen or storm doors provide good curb appeal. For a greater impact, double entry doors can be used, or install a single door with sidelight windows at one or both sides. For a weather seal on double entry doors, a T-shaped wood molding, called an astragal, is installed in the gap between the doors.

doors. There may be rabbets on both edges of the side jambs, one rabbet to stop the primary door and the second rabbet to stop the storm/screen door.

Interior doorjambs are usually made of square-edged stock, with a separate stop molding nailed to the jambs.

Prehung interior door packs are made with jambs to match either wallboard or plaster wall thickness. If your walls are not of standard thickness, you can have the door frames custom-made to fit, or you can add jamb extensions to fur out the jambs. Your dealer can help you match your doorjambs to the thickness of the wall in which they will be installed.

RULE OF THUMB

Exterior Door Sizes

For best access, exterior doors should be at least 32 inches wide (2-8), with 36-inch (3-0) doors preferred.

When finished with ordinary varnishes, exterior wood doors are subject to fading and weather degradation. If you want to show off the wood grain in your door, finish it with an exterior transparent stain. If you like the bright look or sheen of varnish, use spar varnish (used on boats) or a similar exterior product for better durability. Entry doors that are sheltered last longer than doors subjected to direct rain and sunlight. Consider adding a porch or awning to protect wood doors from the weather.

Because higher efficiency became a priority during the energy crisis, steel and fiberglass doors are injected with a core of foam insulation. Magnetic weather stripping reduces air infiltration, so steel entry doors are now more energy efficient than the old door/storm door combination. These efficient doors are a great advantage when installed in tight spaces where an additional storm door would be awkward, such as the entry door between the house and an attached garage, where the steel door also provides a fire barrier between the garage and the house.

To avoid the appearance of a metal flush or slab door, steel doors can have added moldings or be painted or stained with a wood-graining tool to resemble wood doors. Buy the wood-graining tool and finish at your home center or paint store. Directions for wood graining are supplied with the tool. Practice on a scrap of plywood until you master the wood-graining techniques.

Door Security

Most burglars enter the house the same way as the occupants: through the doors. Avoid having entry doors with large areas of glass, because it is easy for a prowler to break the glass and reach in to unlock the door. Ordinary cylindrical or keyed locks can be turned out with a pipe wrench or kicked in by a booted foot. For better door security, install deadbolt locks on all exterior doors. Deadbolt locks have no exterior knobs, so are more difficult to twist out with a wrench. They also have bolts that extend completely through the door frame and into the jack stud alongside the door. These locks are much more difficult to defeat than a simple cylindrical lock.

Door Maintenance

Because they are exposed to the weather and heavier traffic, exterior doors require more frequent maintenance than interior doors. The maintenance directions for both interior and exterior doors is the same.

First, clean and lubricate locks and hinges at least once a year. Ordinary cylindrical or keyed locks are secured in place with four screws: two screws to join the inside and outside doorknobs, and two screws to hold the face of the lock bolt. To clean the locks, take them apart and flush the mechanisms with an aerosol lubricant/penetrant product. Also, door keys gather dirt and lint from being carried in pocket or purse; use an aerosol lubricant to clean grime from door keys.

If the door is sticking, first check to be sure all hinge screws are tight. If a hinge screw is stripped in the screw hole, coat a wooden dowel or wooden golf tee with carpenter's glue, insert it in the screw hole, and use a razor knife to cut off the tee or dowel flush with the jamb. Let the glue set for 24 hours, then drive the hinge screw back into the hole.

If the hinge screws are tight but the wood door is sticking, the door is probably swollen from humidity. You can prevent the door from absorbing humidity by sealing all the door edges, including the top and bottom rail edges. Use an alkyd sealer to seal the door edges when weather is warm and humidity is low.

Check how tightly the lock edge of the door fits against the side jamb. If the door is too tight, sand or plane the lock edge of the door lightly until the door will close. Do not remove too much door stock. If the door is swollen from humidity, there will be too wide a gap between door and jamb when hot, dry weather returns.

One common source of air infiltration is between the bottom door rail and the threshold or sill. Install weather stripping, commonly called a door sweep, to seal the crack between the door and sill. Check door weather stripping and replace it if it is worn. Remember that weather stripping not only seals against air loss but also keeps out dirt and insects.

Patio Doors

Sliding patio doors are available with either aluminum or wood frames. Because aluminum conducts cold and may form condensation or frost in cold weather, aluminum doors should be used in mild climates, while wood doors are superior for use in cold climates. Buy a patio door that is well built, energy efficient, and made with quality materials and hardware. When shopping for a patio door, ask the dealer which door he or she would choose for his or her own home. To ensure good operation and energy efficiency, be sure the door is installed according to the manufacturer's instructions.

Patio doors are now built with improved energy efficiency, both in the glass area and in weather stripping. The door glass may be coated with a low emissivity (low-E) film. On double-pane thermal glass, insulating strips are used to separate the two glass lights, and the cavity between the two lights is filled with insulating gases, such as argon or krypton, rather than dead air.

Look for the listed U-value of any door you choose. The U-value indicates how heat flows through the door unit under laboratory conditions. Door U-values are set by the National Fenestration and Rating Council (NFRC), and the lower the U-value, the better the door. Look for a door with a listed U-value of 0.33.

Patio Door Efficiency

For less air infiltration, consider buying a hinged or swinging door rather than a sliding patio door. Your choice may depend on whether you have enough room to swing the patio door inward. Also, a hinged patio door can offer better security against break-ins than the sliding door in a track.

Garage Doors

Old-style swing-up garage doors were made in one piece, and swung upward to open. Modern garage doors are made in four hinged sections and are called roll-up doors, because they fold and roll up on special door tracks. Roll-up garage doors are available either in single or double door widths, and in a variety of heights depending on the purpose for which they will be used. Garage doors are available in heights from 6 feet up to 12 feet high or more for commercial doors. Before building a garage, consider how high the doors should be: lower for a compact car, higher to accommodate a van.

Garage doors are available in attractive wood designs, or made of steel or fiberglass. If you have no garage door opener, steel and fiberglass doors are lightweight for easy operation. Fiberglass doors may be translucent and let in sunlight.

Insulated garage doors are also available. These doors may be helpful when installed on an attached garage, although the large crack area on a garage door will nullify most of the insulative value. Insulated garage doors are a waste of money when installed on a free-standing garage, unless the garage is heated. (Insulation is intended to stop heat transfer. If there is no heat, no insulation is needed.)

If you have a tuck-under garage, in which the garage is in the basement, under the house living area, you may have limited overhead room for opening the door. For low overhead installations, garage doors are available with a minimum height of only 72 inches (6-0). A low-overhead hardware kit that permits the door to be raised with only 2 inches of top clearance between the door and the garage ceiling can be installed on the door tracks. Shop for the low-overhead kit at home centers. If you can't find the kit in stock, ask your hardware dealer to order it for you.

Garage Door Openers

Garage door openers provide ease of entry, security, and safety. Although there are news reports of injuries caused by openers, in fact hand contact with the door adds to the risk of injury and there are more injuries with garage doors that have no openers. According to the Consumer Products Safety Commission (CPSC), in 1992 there were 17,488 injuries from contact with manual garage doors, and 3,063 reported injuries from garage doors with openers.

Garage door openers are required to have a safety-reverse feature. With safety reverse, the garage door will stop and reverse if it meets a person or object while closing. Check the safety-reverse feature each month to be sure the reverse is working. Place a scrap of wood under the door and close it, watching for it to reverse. If it fails to reverse, call a repairman.

New infrared units can be mounted on the wall on either side of the door. If the infrared beam is broken by an object moving under the door, the opener reverses and raises the door.

Interior Doors

Most interior room-entry doors are made of wood, and are available in widths ranging from 24 inches (2-0) to 36 inches (3-0).

Interior doors are usually hollow-core flush doors, meaning they consist of a frame that is surfaced on both sides by a thin wood veneer. You can buy a door pack, complete with frame, or buy a replacement door only. These doors must be planed to fit into the opening, and the hinges and locks mortised into the door edges and frame. The doors must also be bored for lock holes.

Prehung door packs are easiest for the amateur to install. Prehung packs are shipped with the door hinged and installed in the door frame, and the trim is applied to one side of the door. The lock holes are bored for easy lock installation. After the door is plumbed and shimmed, it is nailed in place and the remaining trim is installed.

Installing a Prehung Door

Prehung doors are sold in what is called a door pack. The prehung pack will include the door frame and doorstop molding, with the door and hinges mounted in place. The trim molding will be installed on one side, with the trim for the other side stapled or tacked to the door frame. The door pack may include a temporary brace tacked to the bottom of the door frame.

To install a prehung door, you will need 4d and 6d finish nails, a nail set, a hammer, tapered wood shims, a handsaw, and a carpenter's level. Remove any temporary braces and trim attached to the door pack. Set the door pack into the rough opening, and use a level to be sure the door is plumb.

Insert the wood shims between the door frame and jamb on the hinged side of the door, at hinge locations. Using the level, check to be sure the door is plumb, then drive 6d finish nails through the jamb and the wood shims. Also insert wood shims at the lock location and the top and bottom of the frame, and check the doorjamb for plumb. Also check the top horizontal trim to be sure the door is level. Nail in place using 6d nails at all shim locations.

Use a handsaw or sharp razor knife to cut off any protruding shims. Then use 4d nails to nail the premitered trim to the jambs.

Cutting Off a Door

When installing new carpeting or finishing an attic or basement, you may need to cut off and shorten a hollow-core door. The problem is that the door has a solid wood frame that extends only about 1½ inches up from the bottom of the door; if you cut off more, you will cut into the hollow core of the door. The solution requires a bit of advanced carpentry, but is doable.

First, mark the bottom of the door where you will cut it off, making a pencil mark on both sides of the door veneer. Next, use a sharp razor knife to cut completely through the veneer on both sides. This will help minimize chipping of the door's face veneer from the circular saw blade. Use wood clamps to clamp a straight cutting guide across the door, then use a circular saw to saw the door off.

Use a sharp wood chisel to shave the wood veneer finish off both sides of the cut-away door frame. The veneer is only glued in place to the bottom of the frame and will come off easily. Now use the sharp chisel to clean the piece of frame and test-fit it back into the bottom hollow of the door. Remove the frame piece, coat it with carpenter's glue, and insert it back into the hollow end of the door. Clamp and let it set until the glue has hardened, then rehang the door.

Saving Space with Doors

A single door that swings from a pair of hinges is the common door type found in older houses. In newer houses, builders use doors that do not take up so much space. In small rooms, consider using double by-pass doors that open by overlapping each other on a track. Other doors for small spaces include bifold doors or pocket doors.

If you are building a closet, consider using louvered doors. Louvered doors permit air to flow into the closet so no musty odors develop. Louvered doors are also preferred for laundry areas or any area that is subject to moisture, such as an entry closet where damp outerwear may be stored.

Doors: Right Hand versus Left Hand

When shopping for a prehung door, you must know whether you need a left-hand or a right-hand door. This refers to the way the door swings when opened.

To determine the swing of a door, face the door so that it swings away from you. Open the door and note which side of the door is hinged. If the door hinges are to your left, it is a left-hand door; if the door hinges are to your right, it is a right-hand door.

RULE OF THUMB

Shimming Doors

Tapered cedar shims are available at home center stores. These shims are inserted to fill the gap between the door frame and jack studs, after the door is plumbed and leveled. Always install wood shims at stress locations on the door, that is, at the lock, corners, and hinges of the door.

Pocket Doors

Rather than swing on hinges, pocket doors (see figure 9.2) slide via an overhead track into a pocket framed into the wall. They are manufactured as a door pack including the frame, door, and metal track. Pocket doors are used in locations where there is no room to swing a hinged door, such as in a bathroom off a narrow hall, or in a bedroom closet.

Closet Doors

Closet doors are made with a height of 80 inches for use in a closet door opening that has a header, or may be 96 inches (8 feet) high for closet doors that reach to the ceiling, with no door header. The 96-inch-high doors have gained popularity because they open the entire closet area, including high shelves, for easy access. Popular closet doors include single doors (which have limited access to storage), pocket doors (which slide into the wall cavity), double doors (two doors that swing in opposite directions), and bifold doors (doors hinged together in pairs that fold away in both directions to open the full door area for access).

Figure 9-2 Pocket Door and Frame

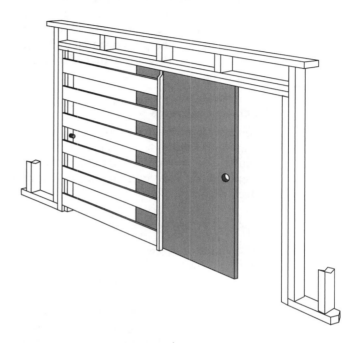

Closet doors are available as flat slab doors, louvered doors, or with raised panels for use in houses that have a Colonial or Early American style.

Installation Tip:
Pocket Doors

Pocket door frames are usually built of 1 × 4 lumber, and these thin framing boards will vibrate from the impact if you try to nail wallboard to them. Instead, use a good construction adhesive or a drywall screw gun and 1-inch-long screws to fasten wallboard to the framing. **Caution: Do not use longer screws that might penetrate through the wallboard and framing and into the pocket door.**

Because closets have little air circulation, they may trap moisture and develop mildew and odors. For improved air circulation and to keep closet contents odor-free, install louvered closet doors. Aromatic cedar chips also help keep stored clothing smelling fresh. Buy aromatic cedar chips at pet supply stores.

Double Doors

Double doors are doors of equal width hinged on both sides of the door frame. Double doors permit access to the full door opening, but may take up needed floor space when opened.

Bifold Doors

Bifold doors are a series of doors, usually each 12 inches wide, that are hinged together and fold back for opening. Two pairs of 12-inch-wide doors are often installed in a 4-foot closet door opening, so that two doors fold away in each direction. Bifold doors can be opened without taking up much floor space, and offer full-door access to the closet.

Bifold doors can be made of wood, steel, or fiberglass. They are mounted via rollers on an overhead track, and may have a floor track to guide the door bottoms.

Windows

Windows are available in a wide variety of styles and sizes. In this chapter we will review the styles of windows available, and will list their advantages and disadvantages.

Because windows are available in a wide variety of standard sizes, we will not cover them all. Your millwork dealer or the window manufacturer will provide you with a catalog listing all their various styles, models, and sizes.

If you are renovating an older house that has windows of odd sizes, don't despair if replacement windows are not available as stock items. Several major window manufacturers will custom-build windows that are exact duplicates of those you are replacing.

Parts of a Window

Before beginning a discussion of windows, the first step is to familiarize yourself with the names of the various window parts.

The glass portion of the window is called the light. Windows may have a single light or, as in Colonial-style windows, may be divided into several lights by wood strips called muntins. Real muntins actually separate a number of smaller lights; false muntins do not divide the window light but simply overlay the light and are for effect only. False muntins are made of plastic or wood, and snap out of the sash for painting or glass cleaning. The sash is the unit in which the glass is set. It may be moved horizontally or vertically, or may swing in or out. The sides or vertical parts are called stiles, and the horizontal top and bottom parts are called rails. The sash is fitted into the window frame.

Window Frame Parts

The exterior trim around a window is called the side casing or the top or header casing. The top of the window is called the header jamb; the vertical sides are called the side jambs; and the bottom of the window is the sill. In a double-hung window, the window sashes are separated by a parting bead; the inside window sash sits against a blind stop.

Window Material Choices

Window frames are made of wood, vinyl, or steel. Wood windows offer good energy efficiency, good appearance, and durability.

Wood windows are available either pre-primed or with exterior vinyl or aluminum cladding. Vinyl windows also can provide good energy efficiency, plus durability and low maintenance. Steel windows offer economy, but because they tend to transmit cold, they are usually used in areas that have mild climates.

Screens and Storm Windows

Older houses may have windows that have screens and storm windows as separate components. The disadvantages of this system are that the older window/storm combination wastes energy and the screens/storms must be changed with the seasons. If your house has this older system, consider replacing the windows with modern efficient units.

Rather than have two separate window/storm window units, modern windows

RULE OF THUMB

Shopping for Window Efficiency

The U-value refers to the rate of heat flow under prescribed conditions through a door or window. The lower the U-value, the more efficient the door or window. When shopping for windows ask the dealer for the National Fenestration and Rating Council (NFRC) U-value for your particular window choice.

have double-pane glass for energy efficiency. The glass may be coated with a low emissivity (low-E) metal film to retard heat loss through the glass. Insulating glass windows have two panes of glass sealed at the perimeter, with a dead-air space of about 3/16 inch between the panes. In modern windows, the space between the two panes of glass may be filled with an energy-conserving gas, such as argon, rather than dead air. Newer windows use the latest weather stripping materials and techniques to achieve low air infiltration.

Modern windows also offer built-in screens. On double-hung windows the screens are installed on the outside of the window frame; on casement or sliding windows the screens are mounted on the inside of the frame.

Window Types

Fixed Sash

Fixed-sash windows do not move in their frames. An example is the picture window. Fixed or picture windows usually have insulating glass, that is, two panes of glass with a 3/16-inch space between them and sealed together at the perimeters. The space is filled with dead air or argon gas. Fixed-sash windows may be installed as a single unit or in tandem with operating windows.

Double Hung

Double-hung windows have upper and lower sashes that move vertically in the frame and when operated bypass each other. They are separated by a parting strip. The windows slide in metal channels and are held in place when open by springs, channel tension, or compression weather stripping. To adjust the tension on the metal channel, check the channels for adjusting screws. Sash locks installed on the meeting rails (i.e., the lower rail of the upper sash and the upper rail of the lower sash) both lock the window sashes tightly together and provide security. The window sashes can be moved either by pulling on metal sash lifts or inserting the fingers into grooves in the sash rails. Double-hung windows can be installed as a single unit or grouped together, either on a flat wall or as sides of a bay window.

Many double-hung sashes can be removed for maintenance by moving each sash to the middle of the window channel and giving a sharp pull on the sash, toward the side with adjustment screws in the channel. The sashes will pop out for painting or glass cleaning. To install, place one end of the sash against the channel on the tension-adjustable side of the window, and push the sash sharply toward that side.

Sliding Windows

Sliding window sashes slide horizontally in tracks mounted on the header jamb and sill. Sliding windows permit opening one side of the slider, but do not open out to catch the wind as casement windows may do.

Casement Windows

Casement window sashes are hinged at one side, and a lever or crank is used to open the window outward. Because of their ease of opening, casement windows are often used over a kitchen sink, where it might be difficult to reach and lift a double-hung window sash.

Awning/Hopper Windows

Awning windows are hinged at the top so the bottom sash opens upward and outward; hopper windows are hinged at the bottom so the top sash swings inward. Awning windows are often installed near ground level, for example, under picture windows. Because they swing outward and

form an awning from the top of the window, they can often be left open even during rain.

Jalousie Windows

Jalousie windows have a series of glass strips that operate via a crank that moves all the strips. The glass strips overlap each other from the top down, much the same way a venetian blind operates. Because there are so many cracks between the glass strips, they leak air and are usually used only in warmer climates.

Bow and Bay Windows

Most window manufacturers offer preconfigured bow and bay windows. The consumer can also group or gang a series of windows together to make a custom bay window. For example, the center window might be either a double-hung or fixed window, with double-hung, casement, or awning windows set at an angle on each side of the center window.

Skylights

If you have a room that is dark, or one that would benefit from increased ventilation, a skylight may be the solution. Although any room may benefit from a skylight, they are especially useful for making a small, dark bathroom seem larger and lighter. In both bathroom and kitchen, a skylight can provide needed ventilation for removing moisture and odors.

Shop at your window supplier for a skylight model that fits your needs. Although most skylights are approved for do-it-yourself installation, the risk of leaks at an improperly installed skylight are high, so we recommend that you have your skylight installed by a professional.

Skylights are usually 4 feet long, and wide enough to fit between two trusses (24 inches) or to span over two trusses (4 feet).

Some have hand cranks for opening, and extension handles are available for reaching high window cranks. Better models have motorized openers that provide easy control from a wall switch.

When installing a skylight on a pitched roof, you will need to build a light tunnel from the window down to the ceiling. To avoid obstructions in the attic/ceiling, you may have to offset the tunnel. The light tunnel must have insulated walls to avoid formation of condensation or frost on the window or the tunnel walls.

Avoiding skylight leaks.   Some skylights have a built-in flashing flange, while others must be flashed with aluminum flashing. If you live in snow country, build a wooden frame or curb around the opening in the roof and mount the skylight on the curb. This will position the skylight so the window is above the top of roof snow. The window can be opened in any weather, and leaks can be avoided.

Sticking Windows

If painted windows stick, suspect that paint has entered into the cracks between the sash and the stops, or between the meeting rails of the upper and lower sashes of a double-hung window. Use an inexpensive serrated knife, called a paint zipper, to cut away paint in window cracks. Just insert the zipper in the crack and move it along the entire length to remove the paint.

To avoid paint runs in window cracks, always work with a dry brush; that is, avoid loading the brush with excess paint. Move the window sash frequently while painting to prevent paint from drying and sticking the sash edges. Never paint the inside surfaces or meeting surfaces of the meeting rails, because paint will cause the surfaces to stick. Instead, leave the inside surfaces of the meeting rails unpainted, and apply only a clear water seal to the bare wood.

Never force a stuck window to open it. Forcing the window can distort the bottom rail and may crack the glass. Instead, use the window zipper to clean the cracks between sashes or sash and stops as directed above. You may have to repeat the procedure several times until you free the sticking sash. If you have crank-out or lever-type casement windows that will not open, first be sure the window is unlocked. Then try the crank or lever again, using only normal pressure. If this fails, push or tap lightly with your hand along the window stile at the crack where the stile and stop meet. This will usually permit you to open the window with no damage to the unit.

Save the window manufacturer's instructions, and clean and lubricate the windows as directed. Wipe window channels or guides with a clean cloth to remove dirt and grime. Vacuum the windowsill area to remove dust.

Updating Old Windows

If your window frames and sashes are in good repair, but the channels and weather stripping have become worn, you can buy window repair kits. These include new channels that will let you eliminate the old sash weight system. To install new window channels, you must carefully remove the interior window trim and stops. Old wood is brittle, so use a small pry bar and pry only at trim nail locations to avoid cracking the wood. Set the trim and stops aside for later reassembly.

With the stops removed, pull out the window sash and cut or unhook any sash ropes. Remove the window sash and inspect it for problems. Do any needed maintenance while you have the sash out of the window. Scrape or sand away any paint runs on the window sash and meeting rails. Examine the glazing compound and replace if it is dried or cracked. Use an electric heat gun or a propane torch to heat and soften the old caulk for removal. Concentrate the heat on the caulk to avoid the possibility of cracking the glass. Repaint the sash. Replace any weather stripping on the window sash.

Remove the old channels and follow the kit instructions to install new channels. When the channels are installed, reassemble the window.

Replacing Windows

In older houses, the windows were made by carpenters on the job, so windows may be odd sizes. As mentioned above, some manufacturers will custom-build windows to fit any opening.

In the post–World War II period, manufacturers standardized window sizes, so you can often find replacement windows of the exact size as the old ones. To find the window size, measure the window across the frame (not including trim), top to bottom and side to side. Take these rough measurements to your dealer or compare the size to catalog models. New windows are shipped completely assembled, with only interior trim to be added.

If you wish to replace a larger window with a smaller one, it is a simple task to add framing to fill in the rough opening. But if you will replace a smaller window with a larger unit, you must remove the exterior siding and sheathing and the interior wallboard or plaster around the new rough opening. Inside the room, install a temporary wall parallel to the window wall to support the roof trusses or joists, and remove the old wall framing out to the size of the new rough opening. Install a longer header that will span the top of the new opening. Install a new stud and cripple stud at each side of the rough opening to support the new header.

To prevent water leaks, staple 8-inch-wide strips of 15-pound asphalt felt around the outside perimeter of the rough opening. Install the side strips first so they extend about 4 inches above and below the header and sill sides of the opening. Staple the top or header felt so it overlaps the top ends of the side strips. Install the window, then staple the bottom strip so it fits tightly against and overlaps the bottom ends of the side strips.

To set the window, first cut off the horns or extensions below the windowsill. Then, using a carpenter's level on the sill, insert cedar shims between the windowsill and the bottom header until the window is level. For wide windows, install shims at intervals to support the length of the sill. When the window is level and plumb, drive galvanized casing nails through the exterior window trim or casing to hold the window in place. Nail at all four corners and at 16-inch intervals around the casing.

Some windows have nailing flanges attached to the frame. Plumb and level the window as above and use galvanized roofing nails to nail through the flanges and into the wall framing.

Window Flashing

To prevent water leaks, install L-shaped metal flashing that overlaps the sheathing and the top or header casing of the window. Then staple an 8-inch-wide strip of 15-pound felt under the bottom of the sill. Insert the ends of the sill flashing under the bottom ends of the side strips, and be sure the top edge of the felt extends up into the groove in the bottom edge of the sill.

Fixing a Broken Window

To replace the glass in a window, use a heat gun or propane torch to soften the old glazing compound, then use a putty knife to scrape away the old glazing. With glazing removed, look for small triangular metal tabs called glazing points, and remove them with the putty knife or screwdriver tip. Wear gloves to remove the broken glass.

To allow for expansion, the replacement glass should be cut ⅛ inch shorter both in length and width than the actual rabbeted opening into which it will fit. Lay a small bead of glazing compound around the rabbeted lip, then gently press the glass into place. Drive new glazing points around the edges of the glass to hold it in place. Then roll a rope of glazing compound about ½ inch in diameter and press it in place over the edges of the glass and against the sash rail. Hold a putty knife at a 45-degree angle to the sash rail, and smooth and remove the excess caulking compound.

Have glass cut at your hardware store, or buy an inexpensive glass cutter. To cut the glass, lay a straightedge along the cut line. Apply mineral spirits on the cut line to lubricate the glass cutter. Use the glass cutter to score along the cut line. To snap the glass, place the cut line over the edge of a workbench or scrap of wood. Press down on the overhanging portion of glass to snap it along the line.

10 Weatherizing and Climate Control

A review of the various energy components of your house can help to minimize your monthly utility costs and keep your household budget under control. An energy-efficient house can also improve your family's comfort level by eliminating air infiltration and drafts. If you have family members who suffer from allergies, asthma, or bronchial complaints, modifications to your present ventilation and heating equipment may help relieve those symptoms. In this chapter we will review insulation, vapor barriers, ventilation, heating and cooling, and humidifiers.

Understanding Home Energy Consumption

Many homeowners have taken the advice of energy experts and upgraded the thermal envelope—the insulation, weather stripping, and windows of the house—only to find that their energy bills did not go down, or in fact went up as energy prices escalated. The problem is that energy conservation does not depend on the energy efficiency of the structure alone.

The amount of energy consumed by the individual family depends on three factors: the thermal efficiency of the building itself (i.e., energy-efficient insulation, weather stripping, and windows); the efficiency of the equipment

and appliances, especially those which provide heating or air-conditioning (furnace, boiler, and air conditioner); and the lifestyle and habits of the family. To reduce energy consumption you must address all three of these factors.

When you have met R-value recommendations for insulation, sealed cracks via weather stripping and caulk, and upgraded your windows to meet modern air infiltration standards, consider upgrading your appliances and equipment. It is most important to inspect the heating and cooling equipment.

Older heating and cooling equipment was highly inefficient, so any appliance that is more than 10 years old is suspect. For example, when energy supplies were both cheap and abundant, forced-air furnaces were built with only 65 percent efficiency. This means that one-third of your energy dollars were sent up the furnace stack. No amount of insulation can offset the fact that your furnace was designed to exhaust one-third of the energy consumed up the stack. Modern furnaces offer as much as 80 to 95 percent efficiency, meaning the furnace delivers from 80 to 95 percent of the fuel energy as usable heat, rather than exhausting it to the exterior.

Air conditioner efficiency is also greatly improved in modern appliances. Consider replacing older, inefficient heating and cooling equipment with modern equipment. In some areas energy utility companies offer bonuses or discounts to the homeowner who replaces inefficient equipment.

One builder's customer survey will illustrate the important part that family lifestyle plays in home energy consumption. In the early 1980s a builder built 20 houses. All the houses had the latest energy-efficient materials and appliances, and were built and insulated to the exact same standards. The builder then proposed that each family record its monthly energy bills, and offered to pay the lowest annual energy bills submitted by the customers. One year later the energy bills of all the families were compared, and the builder was shocked to see that the highest bills submitted were more than double the bills of the most conservative family. The houses and equipment were all cookie-cutter identical, and the only variable was family lifestyle.

What accounts for the lifestyle contribution to high energy costs? Obviously, the more family members, the greater the energy consumption. A houseful of children will require more lighting, more TVs, and more laundry, and children will be running in and out of the house, so air infiltration through the doors is much greater than for a couple or small family.

Family size is not the only criterion. Each degree on the heating or cooling thermostat represents an increase of about 3 percent on your energy bills. Setting the heating thermostat at 73 degrees rather than 68 degrees will increase annual heating bills by 15 percent. Conversely, setting the cooling thermostat at 75 degrees rather than at 80 degrees will increase annual cooling costs by 15 percent or even more. Actually, savings per degree of thermostat change is much greater for air-conditioning equipment than the savings for heating equipment (see "Air-Conditioning" on page 176). As with all of life, there is no free lunch when it comes to energy costs.

Understanding Heat Transfer

A common homeowner mistake is to assume that heat loss occurs primarily or solely through the ceilings. The problem is that how-to texts often confuse principles of heat loss with the fact that hot air rises.

We all know that hot air rises. But heat loss does not occur upward; heat loss is omnidirectional. The rule of thermodynamics is that heat flows to cold, and it will do so through any avenue possible. If you insulate only the ceilings, the heat will escape through the walls and windows; if you insulate both ceilings and walls, the heat will flow outward via the floors or basement

RULE OF THUMB

South-Facing Windows

Each square foot of south-facing window that is insulated at night saves about 1 gallon of oil a year.

walls. Insulation company charts show that in houses with adequate ceiling/wall insulation, the greatest heat loss occurs through the uninsulated basement walls. Insulation achieves its highest efficiency when the insulation blanket is continuous and uninterrupted around the entire exterior perimeter of the house.

The homeowner can often improve the efficiency of insulation by seeking out and filling any small voids in the insulation and by insulating floors or basement walls to provide a continuous barrier to heat loss.

Insulation

Prior to the 1950s no insulation products were available. Ceiling insulation was largely unknown, and weatherproofing the walls was done by back plastering. In this system the exterior sheathing was applied, and the plasterer applied plaster to the sheathing, between the stud cavities. This sealed any cracks between sheathing boards and also sealed the cracks between the framing studs and sheathing. This process greatly reduced air infiltration. When the interior plaster lath and plaster were applied, each stud cavity was effectively sealed both on the interior and exterior sides, so dead air was trapped in the stud cavities, much as the airspace between a vacuum bottle and the outer shell prevents heat loss and keeps coffee hot.

R-values by Region

Recommended insulation R-values vary by region. The R-values are highest for the northern tier of states, and southward through the mountain states. They are only slightly lower through a line drawn roughly through the middle of the country. In the extreme south the temperature swing from winter lows to summer highs is small, and insulation is provided to prevent heat gain

RULE OF THUMB

R-value

The R-value of a given material represents its resistance to heat transfer. This number is used to describe the relative efficiency of insulation. The higher the R-value, the more insulation value a given material has. Any two materials with the same R-value have the same insulation ability. It may take more of a particular insulation material to equal the same R-value in another. R-values give us an easy method to compare different insulation materials.

Recommended R-values are based on the financial point of diminishing returns; that is, the recommended R-value for your area depends on insulation's paying back its cost within a reasonable time. Thus, R-values are arbitrary, and as the cost of fuel goes up, recommended R-values rise.

into the house in warm weather, rather than heat loss from the house interior to outside, as in a colder climate.

Limitations of R-value

R-values are established in laboratories, under carefully controlled, moisture-free conditions. In actual service the R-value or insulation performance is affected by job-site conditions. Insulation that is carelessly installed, with many voids or gaps, will not perform at maximum R-values. The performance of any insulating product diminishes as the discrepancy between interior and exterior temperatures widens; that is, as outdoor temperatures fall, the rate of heat loss rises.

In building products R-value is not the only factor to consider. For example, the logs in log houses do not have a high R-value, yet log houses are energy efficient.

This is because the logs have mass, which increases the ability of the logs to absorb and reflect back interior heat. Logs, therefore, are said to outperform their R-value.

Loose-Fill Insulation

The first available insulation products were loose-fill materials such as vermiculite (R-2.1 per inch), perlite (R-2.6 per inch), and mineral wool (R-3.1 per inch). These products have relatively low R-values but are helpful when poured on ceilings between the ceiling joists. Because these products tend to sag and settle from gravity, they are not efficient when used to fill stud cavities in walls.

Fiberglass Batts and Blankets

Fiberglass is made up of spun glass fibers that can easily be formed into continuous batts or blankets. The batts or blankets can easily be installed between studs or joists. Because the batts have a springy quality, they can be placed between framing members and will expand to hold themselves in place via friction. Fiberglass pellets (R-2.3) are also available for pouring.

Fiberglass batts or blankets have an R-value over three R's per inch of thickness. For example, a batt that is 6½ inches thick will fit between 2 × 6 wall studs, and will provide an insulation value of R-22. When shopping for fiberglass insulation products, just note the listed batt thickness in inches and multiply by 3 to find the total R-value.

Fiberglass is the most popular insulation product for remodeling, retrofitting, or new construction. The advantages include ease of installation, high R-value per inch of thickness, resistance to pest infestation, and fire resistance or failure to support combustion. Fiberglass batts are inexpen-

sive and are widely available to the consumer.

New encapsulated Pink Plus R-25 Miraflex by Owens-Corning is itch-free insulation, because it's wrapped in a polywrap so it's less irritating to touch. The new product is packaged in ultracompact rolls, so it's easy to transport and handle, and slides easily into tight places.

Batts/Blankets: Faced versus Unfaced

Fiberglass batt insulation is available in varying thickness, and either unfaced or faced with a vapor barrier. The vapor barrier may be kraft paper or aluminum, which is adhesively attached to the fiberglass.

The problem with using faced insulation batts is that most installers do not observe the proper installation techniques. In order to provide an efficient vapor barrier, the flanges on the edges of the batts must be carefully overlapped and sealed at each joint. Faced fiberglass batts installed with the edge flanges stapled against the sides of the framing members do not provide a continuous vapor barrier, because moisture may pass between the framing and

RULE OF THUMB

Mixing Insulation

Insulation can be added over existing insulation, regardless of the insulating material used. For example, fiberglass insulation can be installed over existing mineral wool or vermiculite insulation with no adverse results.

the insulation flange. Those flanges must be overlapped on the face of the framing member in order to provide any barrier to vapor penetration. Most builders prefer to install unfaced insulation batts first, then use a continuous polyethylene film as a vapor barrier or retarder.

Pipe/Duct Insulation

Uninsulated pipes and ducts allow heat to escape along their entire length. But energy use is most efficient when the heat is delivered to a specific area, so it is important to insulate pipes and ducts.

In the past asbestos was commonly used for insulating hot water or steam pipes or heating ducts. Because of the danger of health problems, asbestos is now banned for use in building materials, so substitutes have been found for asbestos.

Fiberglass pipe and batt insulation is now available. This insulation comes in rolls that have a facing of aluminum foil. Use duct tape to secure the end of the insulation roll to the pipe, then wrap the insulation around the pipe. For duct insulation, first use duct tape to seal all joints in the ducts. Then fit rigid fiberglass panels around the ducts.

Tubular foam plastic insulation is available for insulating pipes. The insulating tubes are available to fit a variety of pipe diameters. A slot is cut along the length of the tube insulation, and the tube is simply slipped over the pipe and secured at the ends with duct tape.

Vapor Barriers

Vapor barriers (also called vapor retarders) prevent water damage to walls and ceilings by providing a barrier to moisture passage through the building components.

Moisture seeks its own level. Just as water flows downhill to pool in a lake or ocean, when outdoor moisture is lower than the moisture level in the house interior, the indoor moisture will attempt to equalize with outdoor moisture. This attempt of moisture to equalize is a powerful force that causes indoor moisture to pass through walls, floors, and ceilings to reach a balance with outdoor moisture.

The rate of moisture penetration through the walls or ceilings varies with the seasons. As temperatures drop in winter, the relative humidity of outdoor air will drop to desert levels. The indoor humidity levels are augmented by family activities such as bathing and cooking. If indoor moisture levels fall too low, usually below 30 percent, the dry air dries the skin and causes nasal and bronchial irritation. Furniture and house trim may crack or become loose at the joints. In cold climates, homeowners often use humidifiers to augment indoor moisture levels. Thus, the outdoor air is dry and the indoor air is moist, and indoor moisture will pass through walls and ceilings in an attempt to equalize with outdoor moisture levels.

If no barrier is present, moisture in vapor form passes through walls or ceilings. As the vapor moves through building components, it wets insulation, reducing the R-value of the insulation, because moisture conducts cold. Also, if enough moisture vapor is present, framing in the wall may rot or decay. When the vapor reaches the backside of wood siding or trim, the moisture will pass through the wood until it meets the exterior paint. The exterior paint provides a barrier to moisture penetration, so the moisture will lie under the paint film until the sun's rays cause the paint to blister and crack, and the moisture can pass through. Thus, various types of damage can result if moisture is allowed to pass freely through building components. When plastic film sheeting was developed, the builder had available a continuous and efficient vapor barrier.

RULE OF THUMB

Vapor Barriers in Older Houses

Houses built prior to the 1960s do not have a vapor barrier, because none was available. If you own a house with no vapor barrier, you can apply a coat of alkyd paint to the interior surface of ceilings that contain insulation and to the exterior walls to reduce moisture penetration through these exterior surfaces.

Proper Vapor Barrier Installation

The most effective way to establish a vapor barrier is to complete all framing and insulation, then cover all walls and ceilings with continuous sheets of 4-mil (thick) polyethylene sheeting. To perform properly, a vapor barrier must be continuous, with all joints sealed with tape and any accidental holes repaired before application of wallboard. Also use tape to seal around any holes for windows, doors, pipes, or electrical outlets. In many houses the vapor barrier has been carelessly installed, leaving unsealed joints and open holes in the plastic barrier. Moisture transfer is pervasive, and the moisture will pass into walls or ceilings through the smallest opening.

Use a staple gun to attach the plastic vapor barrier to the wall or ceiling. Where possible—for example, when turning a corner from one wall to another—do not cut the plastic sheeting. Wrap the plastic around the corner, stapling as needed to secure the plastic to the surface. Where joints cannot be avoided in the plastic vapor barrier, overlap the plastic by at least 6 inches, then use tape to seal the joint.

Carefully cut out any holes for electrical outlets or pipes and seal around each hole with tape. At windows and doors, leave the plastic barrier intact, overlapping the openings, until wallboard is installed. Nailing the edges of wallboard around window or door openings will hold the plastic barrier in place, while limiting interruption of the plastic barrier. After wallboard is installed (some builders wait until the trim is installed), use a sharp razor knife to carefully cut away the plastic from the window or door openings.

Tyvek

In the past, tar paper was applied over plywood sheathing to form a barrier to infiltration of moisture and air from the outside into the stud cavities in exterior walls. The problem with tar paper is that it forms a vapor barrier between the plywood sheathing and the siding, which blocks moisture movement from the house interior to the outside. Moisture then is trapped in the stud cavities of the exterior walls, or between the sheathing and siding. Tyvek is the brand name of a material that is now used in place of tar paper. Tyvek has the ability to let moisture pass outward, but it provides an air and moisture infiltration barrier that prevents cold air or moisture from passing from the exterior to the interior. Tyvek is often used in conjunction with foam plastic insulating sheathing.

Insulating Sheathing

In the past, wood—either 1 × 8 boards or plywood—was used for house sheathing. Board sheathing is installed at a 45-degree angle to the framing for better strength. Wood sheathing—either board or plywood—provides a nailable surface to which siding can be attached, and also provides bracing strength to the walls and roof.

Various types of foam insulating panels are now available to replace plywood sheathing. These include urethane panels (R-5 to R-8 per inch), polystyrene (R-4.5 per inch), and beadboard (R-3.6 per inch). Foam plastic panels provide insulation value in exterior walls, but the foam panels have no strength, so other bracing, made either of wood or steel, must be installed to reinforce the exterior walls of the house. Siding must be nailed through the studs only, because the foam panels have no nailing strength. Because foam plastic is soft, special large-headed nails must be used to install the foam sheathing.

Foam panel insulating sheathing has an estimated R-value of 3.5 per inch of thickness, and is usually 1 inch thick. But unlike other types of wall insulation, which only fill the wall cavities between the studs and thus provide a reduced R-value through the stud areas, foam insulation provides the same R-value over the entire wall area, so there is no reduced R-value at the studs. This continuous sheathing coverage provides both insulation and an uninterrupted barrier to air infiltration through the walls.

Ventilation

Any room in which family activities produce moisture should be provided with ventilation. The rooms in which excess interior moisture is generated include the kitchen, bath, and laundry. Also, because efficient plastic vapor barriers are now installed in both walls and ceilings, excess interior moisture is now a problem in newer houses. If you have tightened up the exterior of your house by installing more insulation, better weather stripping, and more energy-efficient windows, you may have to install more ventilation to rid the house of excess moisture.

Heat Exchangers

Because of tighter building standards, modern houses permit very little air infiltration, and indoor moisture may build to unacceptable levels. To counter this, many owners of newer houses have been forced to install air-to-air heat exchangers. These heat exchangers consist of two ducts, one for incoming fresh air, the other for outgoing moisture-laden interior air. The two ducts pass through a heat exchanger that transfers the heat from interior air to the incoming fresh air. Thus, fresh air is introduced into the house and moisture is vented to the outside, but very little loss of energy occurs in the transfer process. Heat exchangers are expensive, costing several thousand dollars, but if you have interior moisture problems you cannot solve, consider installing a heat exchanger.

Before installing a heat exchanger, try reducing interior moisture through less expensive means. These options include providing room ventilation to any rooms where moisture is produced, such as the kitchen, bath, and laundry. To exhaust this moisture, be sure to run ventilation fans during or after cooking or bathing.

Kitchen

Ventilation for the kitchen can be provided by through-the-wall or attic exhaust fans, or via ventilation hoods that are installed above the cooktop. Hood ventilators are connected through a duct into the attic, or through the roof; some vent directly through the wall. If you install any ventilating fan, be sure attic ventilation is adequate to exhaust all the moisture pumped into the attic. If you have an older house, extra attic ventilation may be installed to ensure moisture does not build up in the attic insulation.

Bath

Because of family bathing, the bathroom generates much excess moisture. Symptoms of excess bathroom moisture include fogged bath mirrors and mildew buildup. If your bath lacks a ventilation fan, check with your local home center to choose either a through-the-wall or through-the-ceiling ventilation fan.

Attic

In older houses, attic ventilation was woefully inadequate, consisting of louvered gable vents or a couple of static vents installed in the roof deck. Modern attic ventilation is usually continuous soffit-and-ridge venting. These vents are called continuous because they vent the space between each pair of roof rafters, replacing the occasional single vent that once was common both at the soffits and the roof deck.

Continuous soffit vents permit outdoor air to be pulled into the attic between each pair of roof rafters or trusses and to flow across the attic insulation. As the air passes through the attic, it picks up moisture, is warmed, and rises to exit via a continuous ridge vent. This natural action, in which dry outdoor air is pulled in to replace the warm, moist attic air exhausted via the ridge vents, is called the chimney effect, so called because the rising air movement is provided by the warmed rising air.

One way to check whether your attic vents are adequate is to check your attic for signs of moisture buildup. If insulation feels damp or there are water stains on the underside of the sheathing, you may have to upgrade your attic ventilation. If you live in a cold climate, check your attic during very cold weather, looking for frost buildup on roof rafters or sheathing. If you see any frost, you may need more vent capacity.

To determine whether attic ventilation is adequate, have your roof and attic inspected by a roofing contractor. The best time to upgrade attic ventilation is when you reroof. At that time the contractor can install continuous soffit-and-ridge vents.

Whole-House Fans

Whole-house fans were introduced in southern states before central air-conditioning was available. The theory is that a large central fan, usually mounted in the ceiling of a center hall, could be used to cool and ventilate the house. Choosing the right size whole-house fan is important, for it must be able to move enough air through your house to be effective in providing a cooling benefit. Figure 10.1 shows how whole-house fans move air through the house when installed either in the ceiling or a gable wall.

The Home Ventilating Institute has an easy method to calculate the size of whole-house fan you will need to effectively cool your house. All you have to do is calculate the square footage of the living space in your house and multiply this number by 3. For example, if you have a 1,500-square-foot house, you would need a fan that is rated to move at least 4,500 cubic feet of air a minute. If you live in a warm climate, multiply the living area by 4.

For most efficient cooling with a whole-house fan, open the windows during the coolest part of the evening. The coolest air will be at ground level, so it is important that the lowest windows in the house be opened. Turn on the fan. It will pull the cool night air through the entire house and exhaust it into the attic. The night air will cool both the interior air and the contents of the house. The interior temperature of the house will then be equal to the overnight outdoor low temperature, perhaps 70 degrees.

At sunrise, close the windows and avoid opening windows or doors during the

Figure 10-1 Whole-House Fan Installation

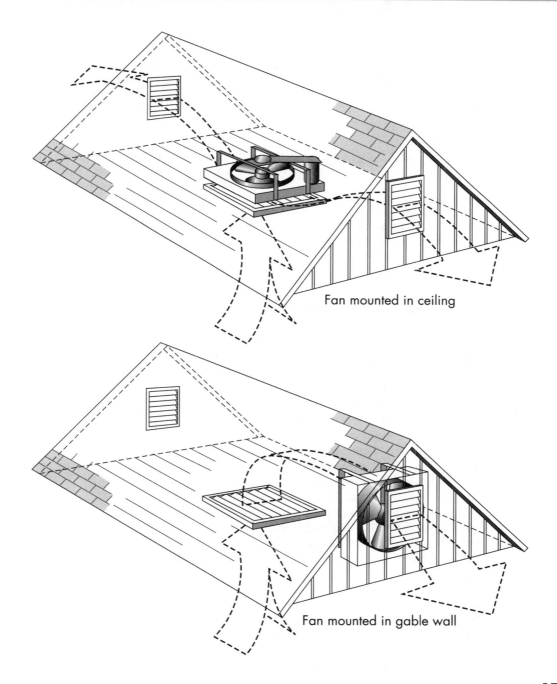

Fan mounted in ceiling

Fan mounted in gable wall

RULE OF THUMB

Fan Size

Fan dealers often sell small-capacity fans as whole-house fans. To cool the whole house, a fan must be large enough to produce one air change per minute. A fan of lower capacity will not do an adequate cooling job.

daytime hours. Also close curtains or drapes to limit solar heat gain into the house. As the outdoor temperature rises, the interior temperature will rise only about 1 degree per hour.

Crawl Space Ventilation

If your house is built over a crawl space rather than a basement, that crawl space may require ventilation. Modern construction techniques can produce a crawl space that needs no ventilation. To achieve a vent-free crawl space, the house must be built upon well-drained soil, and the soil must be covered with a vapor barrier film of poly-ethylene plastic sheeting, at least 6 mil thick. The vapor barrier should be folded up about 6 inches along each foundation wall, and sealed to the foundation with mastic. Then 3 inches of sand should be spread over the entire area enclosed by the foundation. This approach can be used to help stop moisture problems in an existing crawl space.

Vents are required in crawl spaces that were not constructed to be moisture-free. If you have a crawl space that has vents, go into the space and check the underside of the floor with a strong light. Look for any signs of moisture damage, or wet spots on

the ground or the underside of the flooring. If you see signs of wood rot or other moisture damage, you should install more vents.

It is a common mistake to automatically close off crawl space vents in winter. Keeping floors warm must be done by proper insulation, not by eliminating ventilation. The underside of the floor should be well insulated to keep the floors warm, and good ventilation should be provided to exhaust winter moisture. The decision of how much and when to ventilate the space should be made after seasonal inspection of the space. If no moisture problems are evident, you can close the vents in winter. If moisture is present, open the vents in winter.

Heating and Cooling

The most expensive energy cost is for interior heating or cooling. Having the most efficient heating and cooling equipment can both increase family comfort and reduce energy costs. If your heating or cooling equipment is more than 10 years old, consider replacing it with modern efficient equipment.

Heating

Fuels

In the United States, the most common type of home heating furnace is the forced-air furnace, and the most common fuel is natural gas. Where natural gas is not available, heating with oil or propane becomes the next option. In most markets heating with oil is more expensive than heating with natural gas, while propane gas is the most expensive of the three fuels. Because natural gas is abundant and can be used to fuel other appliances, such as cookstoves, water heaters, or clothes dryers, natural gas should be the fuel of choice where it is available.

Hot-water baseboard heat provides a very even heat without the drafts associated with forced-air equipment. Radiant hot-water equipment can be fired with either natural gas or heating oil.

One heating choice that is becoming more popular is in-floor radiant heating. Architect Frank Lloyd Wright pronounced in-floor heating to be the best heating system available. If you are building new, check out the in-floor radiant heating systems available.

For a time in the 1970s, electric power companies promoted electrical radiant heating. This type of heating may be feasible in moderate climates, but has proven too expensive for use in cold climates.

Space Heaters

Various types of space heaters are available for heating limited areas, or for providing an auxiliary heat source in cold weather. These heaters may be fueled by oil, kerosene, natural gas, electricity, or wood.

Space heaters that use electricity vary from direct radiant models to ceramic models. Because they are combustion-free, they give off no dangerous gases and need no venting. They can be thermostatically controlled, and feature safety switches that turn the units off if they tip over.

At the height of the energy crunch, space heaters fired by fossil fuels and requiring no venting were widely advertised. But the energy crisis that caused a boom in space heater sales also brought about tougher building standards to tighten up houses against air infiltration, and air pollution experts agree that any combustion-type space heater that uses fossil fuels should be vented to avoid buildup of dangerous gases such as carbon monoxide. All fossil fuels produce gases as a by-product of combustion. Even gas cookstoves, which are exempt from venting requirements on the theory that they are used only occasionally and for short periods, can produce high levels of exhaust gases when used to cook a holiday meal. When using a gas range for an extended period, open a door or window, or run the exhaust hood fan.

For many of us, burning wood in a space heater or fireplace conjures up nostalgic memories. The cheery open flame, comforting warmth, and aroma of burning wood are all fond winter memories. But surveys have shown that children who live in houses where wood is burned suffer far more colds and bronchial distress than those who do not. If you burn wood in a space heater, install a proper chimney, burn only seasoned wood, and clean the stove and chimney regularly to ensure a clean burn and to minimize air contamination.

Fireplaces

Masonry fireplaces. Masonry fireplaces are built of brick or stone, and require heavy footings to support the massive weight. Both because of the cost of materials and the amount of labor required to build them, masonry fireplaces are expensive. Because hot air rises, and a fireplace chimney is essentially a hole in the ceiling, masonry fireplaces can be energy wasters. They exhaust not only the heat and smoke they produce, but also draw out air that has been heated by the furnace. If you have a masonry fireplace, install ducts that pull in combustion air from outside, and install a fireplace insert or well-fitted glass doors to prevent conditioned air from being lost up the chimney. When no fire is burning, close the damper and the glass doors to block the air exit.

Zero-clearance fireplaces. Both to combat rising construction costs and to build fireplaces that are more energy efficient, fabricated fireplace units have been developed. They are called zero-clearance fireplaces because, when proper installation techniques are observed, they can be installed without fire danger over a wood floor or against a frame wall. Zero-clearance units are relatively inexpensive and lightweight, require no expensive footings, and can be installed using a metal chimney. A wood-framed enclosure can be built to cover the fireplace face, and can be finished with wallboard or man-made brick or stone. The metal chimney can be concealed by a ply-wood covering or chase, which also can be covered with brick or stone.

Zero-clearance fireplaces can be installed to draw combustion air from outside. Both wood- and gas-burning models are available. Visit a fireplace dealer to select the right model and to obtain complete do-it-yourself installation instructions.

Cooling

Air-Conditioning

Central air-conditioning. Aside from natural cooling, central air-conditioning is the most efficient way to cool your entire home. If you have a forced-air furnace, you already have the air distribution ducts to carry cooled air throughout the house. Shop carefully, and buy the air conditioner with the highest energy efficiency rating (EER). To install central air, position the condenser in the furnace air plenum, and set the compressor outdoors on a poured concrete slab.

If you do not keep the unit clean, it becomes a dust distributor through the house. For proper operation of both furnace and air conditioner, you must have the ducts cleaned at least every 2 years; remove and clean the backside of room registers at least annually.

Basic air-conditioner maintenance is simple. Replace the furnace filter monthly when either the furnace or air conditioner is in use. Ordinary fiberglass filters are mounted in the cold-air return, filtering out dust before the return air enters the furnace. These filters are called dust-stop filters, and filter out only particles of dust that are larger than 100 microns in diameter. These filters cost about $1 each.

More efficient filters, called media air filters, trap particles of lint, dust, or pollen as small as 0.5 micron. If a family member has allergies, try using a media air filter. These filters may cost between $8 and $10, and are available from your heating supply dealer.

For the maximum in clean air, install an electronic filter. Some electronic filters can cost $1,000 or more installed, but if your doctor orders this improvement for family health reasons, you may be able to take the electronic filter as a medical deduction on your income taxes.

Use a screwdriver to remove the top of your outdoor compressor unit. Use a hose and nozzle to clean dust and dirt from the louvered compressor cover. Check the owner's manual or visually inspect the compressor motor to see if it has oiling ports, and use 30-weight oil to lubricate the motor each spring or as the manufacturer suggests.

To reduce air-conditioning costs, install awnings to shade windows on the sunny side(s) of the house, and use drapes or shutters on the interior side of the windows. Remember that the unit has to work harder if you set the thermostat lower. Whereas lowering the thermostat setting for heating will reduce energy costs only about 3 percent per degree of temperature difference, raising the thermostat setting for air-conditioning by a mere 3 degrees may reduce cooling bills by as much as one-third.

Calculating EER

To figure the EER (energy efficiency rating) of an appliance, divide the BTU output of the unit by the wattage. The formula is BTU/watts = EER.

Room air conditioners. Room air conditioners can be installed either in window openings or through the wall, and can be used to cool single rooms or other small areas. As with any appliance, for best efficiency you should shop for the highest EER available. When shopping, take along the room measurements and ask your dealer to help calculate the size of air conditioner you need. More efficient units consume less energy and can be plugged into any 15-amp outlet; power-hungry larger units may require that you install a dedicated 220-amp power outlet. For safety, be sure you plug the unit into a three-prong grounded outlet.

Use only heavy-duty air-conditioner extension cords. The best policy is to add an outlet at the air-conditioner location to eliminate the need for an extension cord to carry power to the unit.

Room air conditioners are easy to maintain. Keep the filters clean; many units have washable filters. Condenser fins on the outside of the unit may become grimy, and these fins can be vacuumed. If the grime will not yield to the vacuum, ask a serviceman to steam-clean the unit. Because an air conditioner also removes humidity, it must dispose of the collected water. Slightly tilt the air conditioner to the outside so the water will drain away from the room side. Clean and inspect condensate drainage holes or pipes to be sure they are not blocked and water can exit the unit.

Window/Ceiling Fans

By managing your energy picture carefully, you may be able to be comfortable while using the air conditioner on only the warmest days. Often the problem may be high humidity rather than high temperature. It is cheaper to use a dehumidifier to remove the humidity, or to set the air conditioner on low. Also, a variety of fans are available that can help maintain comfort at a lower cost than air-conditioning.

The most efficient fans are whole-house fans (see "Whole-House Fans" earlier in this chapter). These are high-capacity fans that pull outdoor air into the house during the coolest part of the evening, and exhaust the hot interior air.

We are more comfortable when moving air helps evaporate moisture from the skin. Portable or window fans cool you by moving the air. Plan fan placement, using the fan to pull outside air in through the window in one room and exhausting it through an open window in another room. In this way the fan can help cool a larger area.

Ceiling fans. Like other fans, a ceiling fan can provide personal comfort at low cost. Ceiling fans do not pull in outside air, but instead circulate the air within a single room.

Most smaller room fans can be installed at an existing ceiling light box. The light box may be strong enough to support the weight of a small fan, but when installing a larger, heavier fan it may be necessary to provide additional support to the light box. For safety, hire an electrician to install large or heavy fans.

It is a mistake to use a ceiling fan in winter in an attempt to destratify or pull warm air down from the ceiling. Actually, the moving air will make you feel cooler, and to pull warm air downward you must use a fan that has a duct to direct the air downward.

Humidifiers

In cold weather the relative humidity of cold outdoor air may drop, with a corresponding drop in indoor humidity levels. Indoor humidity levels in winter may drop to as little as 15 percent, about the level of humidity in the Sahara Desert. The superdried interior air can cause wood floors to shrink and cracks to open, open large cracks in trim miter joints, loosen legs on furniture, increase static electricity levels, and cause nasal and respiratory problems to residents. To increase indoor humidity levels, use a humidifier.

Humidifiers can be installed in furnace ductwork, or they can be portable units. Calculate the cubic footage of your house interior (floor space × ceiling height = cubic feet), and ask your dealer to help you choose the proper unit for your needs.

Recommended interior humidity levels are around 35 percent. Quality humidifiers will have an automatic humidistat that will turn the unit on or off when preset humidity levels are reached. If you find that humidity levels are too high, check your windowpanes on a cold day for fogging or condensation. Then turn down the humidity level until the condensation on the windowpanes is eliminated.

Humidifiers that become dirty can be a breeding ground for various molds and fungi. Use only clean water in your humidifier, and clean the unit often. Use a recommended cleaning product to kill mold spores. Check your owner's manual for proper cleaning techniques.

11 Exterior

The exterior components of the house are exposed to the damaging effects of spring storms, blistering summers, and, in much of the country, frigid winters. It is small wonder that exterior components such as roof shingles, siding, soffits, and decks fade, peel, crack, and decay. The good news is that when you replace expensive components today, you can choose those which are virtually decay- and even maintenance-free. When replacing exterior materials, remember that it does not cost any more in labor to install a shingle with a 35-year warranty than to install a cheaper shingle with a 20- or 25-year warranty. By the same token, permanent, maintenance-free siding materials cost a bit more than bare wood products, but you are nailing the "paint" on. These products are prefinished and eliminate both the initial painting job and also the need to repaint on a 5- to 7-year cycle. In fact, building materials have become so durable that you can build a house today and know that you will spend no money for maintenance or replacement during the life of a 30-year mortgage.

Gutters/ Downspouts

To understand the need for a roof gutter system, consider that a 1-inch rain will deposit 1,500 gallons of water on a roof that is 2,400 square feet. If this amount of water is allowed to run off the roof edge and collect around the house foundation, the house can suffer from water damage ranging from a wet basement to high indoor humidity, faded siding, peeling paint, and ruined insulation. The gutter and downspout system on your house is designed to catch the water on your roof, deliver it away from the foundation or basement, and deposit it harmlessly onto the lawn or into storm sewer drains. When you decide to install gutters, choose an experienced contractor who can build a water delivery system that will protect your house in even the worst possible storm.

Do It Yourself versus Hiring a Pro to Install Gutters

Before you decide to install gutters yourself, for economy, be sure you are considering all the costs of the project.

Be aware that small items such as gutter hangers and elbows can add up, so estimate the cost of all the components before you reject a contractor's bid.

If your house has no rain gutters, do not assume that the house does not need them. The decision on whether to install gutters or not was made by a contractor (perhaps simply trying to save money) or by a former owner. If the basement is damp or water pools near the house, install gutters.

Basic Parts of a Roof Gutter System

The parts of a roof gutter system are shown in figure 11.1. The basic parts include the gutters that actually collect the rain runoff from the roof and are located along the roof edge or fascia. The gutter sections are held in place with fascia brackets or strap hangers. The straight sections of gutter are joined together with slip-joint connectors. Inside and outside corner-miter pieces connect gutter sections at corners. The water is contained inside the gutter system by end caps. End pieces with downspout outlets connect the downspouts to the gutters, which carry the water to ground level. Several styles of elbows provide for flexibility in downspout placement. The downspouts are secured to the building with straps. Water is diverted away from the foundation with shoe fittings that attach to the bottom of the downspout. Other parts include gutter spikes or hangers to secure the gutters to the roof edge or fascia and strainer caps to keep leaves out of the downspouts.

Galvanized Steel Gutters

Until the 1960s, when aluminum and plastic gutters began to appear, most residential rain gutters were made of galvanized steel.

Advantages of galvanized steel gutters are that they are inexpensive to buy and quite durable, often lasting for 40 to 50 years if properly installed. Steel gutters are coated with a layer of zinc, or galvanizing, that helps protect the steel from rust. But galvanized steel is rust-resistant, not rust-proof, and in time the gutters can rust out and leak.

Another problem with galvanized steel gutters is that when painted they almost invariably peel. This peeling problem is the fault of the paint and/or the painter, not the gutters. Thick house paint will peel on gutters. Metal exposed to the elements of sunlight and cold expand and contract at a high rate, and it is the thick paint coating, applied to the expanding metal, that loosens the paint film and causes the gutters to peel.

Painting Galvanized Steel Gutters

If galvanized gutters are peeling, use a heat gun or paint stripper to remove the old paint. Recoat with an oil-based paint product formulated for metal, and thin the paint to the maximum recommended by the manufacturer. The thinner paint film will not crack and peel as readily as a thick coat of house paint.

Installation of Galvanized Steel Gutters

Galvanized steel gutters are sold in 10-foot-long sections. To join the sections, solder them together at the ends. It is easiest to solder the sections together at ground level, then lift the length of gutter into place. Use your choice of metal gutter brackets or gutter spikes to hold the gutters in place. Bear in mind that gutters must be installed so that they slope slightly to the ends or corners of the house, where the downspouts are. Steel gutters that sag between brackets allow water to pool and to prematurely rust through the steel.

Figure 11-1 Roof Gutter System

Gutter
comes in 10' length

End Piece with Outlet
used where downspout
connects

Downspout
comes in 10' lengths

Inside Miter
used for inside turn
in gutter

Outside Miter
used for outside turn
in gutter

Slip Joint Connector
used to connect joints
of gutter

Strainer Cap
slips over outlet in end piece
as a strainer

End Caps with Outlet
used at ends of
gutter runs

Fascia Bracket
used to hold gutter
to fascia on wall

Elbow Style A
for diverting downspout
to left or right

Elbow Style B
for diverting downspout
in or out from wall

Shoe
used to throw water
to splasherblock

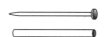

Spike and Ferrule
used to hold gutter
to eave of roof

Strap Hanger
connects to eave of roof
to hold gutter

Connector Pipe or Clincher
used to hold downspout
securely to wall

Aluminum Gutters

Aluminum gutters do not rust and are pre-finished at the factory, so they need no painting. They are available in 10-foot lengths, and are joined by special connectors. Alternatively, a contractor who installs aluminum gutters will shape them at the job site, and install them in seamless lengths so there are no joints to leak or fail.

Plastic Gutters

Plastic gutters are usually available in either white or brown. They are relatively inexpensive and are advertised as a DIY item for homeowners. Plastic gutters never rust, never need painting, and in many cases snap together via special connectors and brackets. Plastic gutters may sag if gutter supports are spaced too far apart.

Plastic gutters are aimed at the DIY market, and installation is simple. Components snap together at connectors, and gutter hangers are simply screwed to the fascia; then the gutters are snapped into grooves in the hangers.

Because plastic gutters do not rust, they need little maintenance. If you live in a snow region, check any rain gutter in the spring, to inspect for damage from roof ice dams.

Wood Gutters

Wood gutters were once commonly used on Early American and Victorian houses. The gutters were simply built into the roof soffit-fascia area. When built of rot-resistant wood like cedar or redwood, or when the inside of the gutters was coated with tar, the wood gutters proved very durable. Wood gutters are not commonly available today, because metal and plastic gutters offer both low cost and durability. If you wish to use wood gutters on a house today, to preserve

Gutter Maintenance Failures

The most commonly neglected gutter maintenance problems are resetting gutters that have been loosened by storms, having sufficiently long ground pipes to carry water away from the house, and keeping gutters clean. Be aware that a half dozen leaves can clog the gutters at the downspouts, so constant inspection is required.

the architectural and historical integrity of a vintage house, you must have the wood gutters custom-made.

Copper Gutters

Copper gutters are strictly top of the line in both cost and durability. Because of their high cost they are seldom used on any but mansion-class houses. Copper resists corrosion and will age to an attractive light green patina.

Roofing

Roofing contractors compare the roof to a good insurance policy: it covers everything. That is literally true. Every component of a house is protected by the roof. Roof leaks can damage siding, windows, insulation, and all the interior finish materials. A neglected roof can result in thousands of dollars of damage to a house, so the homeowner should be aware of the remaining life of the roof, and inspect it for damage each spring and fall, and after any storm activity.

How to Inspect a Roof

When conducting an inspection of your roof, or when buying a house, you can gain a good idea of the roof's condition by a visual inspection. Your roof inspection must be done both outside and inside the house.

Outside Inspection of Roof

- You need not climb on the roof to inspect it. If climbing is a problem for you, or the roof is too steep for safety, use a pair of binoculars to inspect the roof from ground level.

- About 90 percent of residential roofs are asphalt shingles. If the shingles are asphalt, look for any shingles that are windblown, cracked, curled, or broken. The sun's rays are the chief cause of wear on asphalt shingles. As the oils dry out, the shingles deteriorate, which shows up as cracking or curling of the shingles. As they age, the shingles also begin to shed their protective granules. Check in the rain gutters, on splasherblocks (at the base of the downspout, used to direct water away from the foundation), or on the ground under the downspouts for any accumulation of granules.

- Inspect the metal flashing around fireplace chimneys or vent stacks through the roof. In older houses the flashing is often galvanized sheet metal. The metal flashings rust in time, and badly rusted flashing indicates that the flashing may be leaking, or may develop leaks in the near future.

Inside Inspection of Roof

- Check for water stains on ceilings, particularly along exterior walls or on ceilings where they meet a fireplace chimney.

- Check also for any signs of recent ceiling repairs, a clue that the seller of a house you're inspecting may have done cosmetic repairs to cover up a roof leak problem.

- Check the attic, and take along a bright trouble light (portable work light). Do a thorough inspection of the underside of the roof (sheathing), checking for any water stains.

- While you're in the attic, check closely around any vents or chimneys that penetrate through the roof, again looking for water stains, rust, or any sign of water entry.

- Check your household records and try to determine the age of the roof shingles. In past decades asphalt shingles had a felt backing, and carried only 20-year warranties. If the shingles are 20 years old, or older, you will soon have to replace the roof.

Wood Shingle Inspection

If your roof is wood shingles or shakes, look for any signs of deterioration of the wood.

- Check the keyways between the shingles to be sure they are not clogged with debris that can block water and cause it to dam up behind the shingles.

- Check also for any moss growth on the shingles. Moss grows where shingles are shaded or stay damp long after a rain. This constant dampness will cause premature loss of wood shingles or shakes.

Asphalt Shingles

As stated earlier, about 90 percent of residential roofs are asphalt shingles. The first asphalt shingles had a life expectancy of only 15 to 20 years, but in the past two decades there have been great improvements in asphalt shingle construction. The old shingle granules were gravel, and tended to fail early. Modern shingles have ceramic granules that provide better shingle protection and wear than gravel granules.

Another improvement in shingle quality is due to the use of fiberglass mats to replace the old felt mats. Fiberglass mats are far more durable, and these advances in shingle construction have led to shingles that carry manufacturers' warranties of 35 years. When you reroof, choose the best-quality shingle—one that offers a 35-year warranty.

Shingle Coverage

Shingles are measured, sold, and installed by the square; a square is 100 square feet of roof area. Asphalt shingles are sold by the bundle; a bundle is equal to one-third of a square, or enough shingles to cover $33\frac{1}{3}$ square feet. Thus, three bundles of shingles will cover one square or 100 square feet of roof.

To check how many square feet or squares your roof covers, follow this formula. The method applies to a gable roof (a roof with a gable at each end of the house). You must use a bit more advanced math to figure the roof size if the house is L-shaped, or if the roof is a hip roof (a roof that has no gable ends, but extends from the ridge board at the center of the roof down to the eaves, on all four sides). First, find the width of the house. Let us assume a house that is 26 feet wide, or 15 feet (including a 2-foot-wide soffit, or roof overhang) from the edge of the roof trim board, or fascia, to the center of the house. Now find the rise

of the roof. The rise is equal to the distance from the attic floor to the top of the ridge board, and can be measured from the attic. Let us assume a rise of 4 feet. To find the run—the length of the roof from the center of the top of the ridge board to the far end of the roof, at the roof edge or fascia—consider that you are figuring the hypotenuse of a right triangle, with the missing run measurement being the hypotenuse. To find the run of the roof, use the following formula: $a^2 + b^2 = c^2$ with c being the hypotenuse of the triangle and therefore the run of the roof. Thus, we have $15 \times 15 + 4 \times 4 = 241$; the square root of 241 is $15\frac{1}{2}$ feet. So now we know that the run of the roof is $15\frac{1}{2}$ feet.

Measure the entire length of the house, plus any soffit or overhang at the gable ends. Let us assume the house is 48 feet long. Multiply the length times the run of the roof: $48 \times 15.5 = 744$ square feet. Multiply this times 2 (to get both halves of the roof): $744 \times 2 = 1,488$ square feet, or just under 15 squares of roofing. To estimate the material costs (shingles only), check the per-bundle price of shingles at your building center. There will be other costs, including new flashing, drip edge, nails, roof mastic, and the like. Your supplier can help you estimate the cost of these items.

Installing a New Roof

Shingle manufacturers once were reluctant to recommend roofing as a DIY job. Today, asphalt shingles are sold in every home center, and instructions for shingle installation are often available from the manufacturer. But most professional roofers still regard roofing to be beyond the skills and abilities of all but the most skilled DIYers. Why so? First, reroofing time is the best time to replace damaged flashing and valleys, and to upgrade attic ventilation to modern standards. The sheathing should be inspected, and rotted or damaged sheathing should be replaced. These are jobs for a professional. Also, climbing requires that the roofer be in good physical condition, as do kneeling and carrying the shingles. Ladders and scaffolding are needed for the job. (The type and amount needed depend on the type of roof.) So count the rental costs of scaffolding in when you estimate any possible savings from DIY.

Loose shingle scraps become skateboards when you step on them; they can cause you to slide off the roof. To avoid falls, professional roofers keep the roof clean. The professional roofer may tie a rope around his waist, and anchor it as a lifeline to a chimney. Special ladder-type footholds are necessary to provide safety on steep roofs. For these and other reasons, reroofing a house is a difficult job. Most roofs should be done by professionals. If you decide to undertake it yourself, remember the roofer's lament: Never step back to admire your work.

Cost of Professional Installation

The cost of professional roofing varies with the quality and cost of the shingle you select, the condition of the existing roof, whether the roofer must do a teardown (removal and disposal of the old shingles), and the cost of building in your particular area. As mentioned, you can easily learn the cost of the shingles by checking bundle prices at your local home center. To find a contractor, the best approach is to check with your supplier, who often will recommend various subcontractors. Get several bids, from established local firms, and check the contractors' references, including bank references, customer references, and the Better Business Bureau.

RULE OF THUMB

Roof Layers

In most cities, building code allows you to roof over an existing one-layer roof, but you must tear off and start new from the sheathing if the roof already has two layers of shingles. Check your local codes before roofing.

Roll Roofing

A roof that has a slope of less than 4 inches per running foot should not be shingled, because on a low-slope roof, water may run back under the shingles and cause a roof leak.

Because a low-slope roof is less dangerous to work on than a steep roof, many DIYers tackle the job of roll roofing. The first step is to strip off any existing roofing. Then lay a first course of 15-pound felt over the roof. Overlap the joints at least 6 inches, and use a hammer/stapler to fasten the felt underlayment in place.

After laying two courses of shingles, or one course of roll roofing, use a chalk line to guide the alignment of the bottom edge of the next course of roofing. This will ensure that roofing courses stay straight and present a good appearance.

RULE OF THUMB

When to Use Roll Roofing

Look at the roof from ground level. If the slope is so slight that you cannot see the roofing from the ground, use roll roofing.

Next, apply the roll roofing. Mineral-surface roll roofing has granules resembling the granules on asphalt shingles. It is available in weights of 55 to 90 pounds per roll. A roll of roofing is 36 inches wide and 36 or 38 feet long, and will cover one square (100 square feet) per roll when applied in single coverage.

For single coverage the overlap at end joints is 6 inches, and top lap is 2 to 4 inches. A 9-inch strip of roofing is first applied at the eaves, then full-wide (36-inch) courses are laid up the roof. The roll roofing can be installed with exposed nails along the edges or with concealed nails. With concealed nail application, only the upper edges of the roofing are nailed, at 4-inch intervals. The roofing is then overlapped 2 to 4 inches at the bottom edge. When all the roofing is laid, the bottom edge of each course is lifted and lap cement is applied to the lap area; then the edges of the courses are pressed into contact with the lap cement.

For a more durable roof, an alternative is to double-cover the roof. Because of the possibility of hail damage, and to provide superior insurance against leaks, most roll roofing should be applied in double coverage. Any end joints are overlapped 6 inches; in double-coverage application, the top lap or overlap is 19 inches, leaving 17 inches of exposure per course. Apply lap cement to the entire 19-inch-wide overlap area. Double coverage results in having two layers of roofing instead of a single layer over the entire roof. If you use the double-coverage method, you will use two rolls of roofing per square instead of the single roll used for a single layer.

Roofing should be applied only when temperatures are above 45 degrees Fahrenheit. Store lap cement in a warm place. If the cement is too thick, do not heat it over a fire. Because it may be flammable, place the pail of cement in a container of hot water to bring it to working consistency.

Metal Roofing

Metal roofing consists of either metal shingles (e.g., aluminum) or metal panels. Because they do not corrode, aluminum shingles offer long life. Aluminum-shingle roofs also provide protection against ice dams in cold climates. Any snow load will slide off the roof in a miniavalanche when sunshine warms the roof, so the snow does not lie on the roof to melt. Aluminum shingles may offer warranties of 50 years, but are more expensive than asphalt shingles.

Roofs consisting of metal panels with edges that lock together are sometimes used to roof more expensive houses. They offer almost complete freedom from maintenance, but the initial cost is beyond the budget of most buyers.

Roofing Nails

Roofing nails are usually galvanized (zinc-coated) to prevent rusting. Aluminum nails may also be used. They may have a barbed shank for better holding power. They have a large head to hold the shingle more securely against wind damage. In high-wind areas you should also install shingles with a self-sealing strip. The strip of mastic on the shingles will soften in the sun and cause the shingles to bond together for better windproofing.

Some roofing contractors use pneumatic staple guns to fasten shingles with staples. Most old-timers prefer the nail method, claiming the shingle nails have superior holding power compared to staples. When hiring a professional roofer, specify whether you want nails or staples on your roof.

Siding

Except for stucco, brick, or stone, all exterior wall finishes are called siding. Traditional siding is cut from wood in various shapes and species. Within the past two decades other man-made sidings have emerged. These man-made siding materials include plywood or hardboard, steel, aluminum, and vinyl. These siding materials are available in a wide variety of styles and finishes. As labor and interest costs have escalated, prefinished siding that needs no painting and has limited future maintenance has become more popular.

Siding: Do It Yourself or Pro?

Like roofing, siding a house requires a high degree of knowledge. Siding must be laid out so the courses meet and are level to one another at all corners of the house. To get a weatherproof application, lap siding should be installed so that courses fall even with the tops and bottoms of windows.

You must know how to lay out and use a story pole, a pole that is marked where each course will fall, to use as an installation guide. For outside corners on shingle or shake siding, you must work the siding so that the shingles lap on alternate courses of shingles. For lap siding there are choices to make in the way you handle both inside and outside corners. To allow for expansion and contraction as temperatures change, metal siding nails should be driven so they are slightly loose against the flange.

In short, siding installation is an education in itself. Many of the prefinished siding products are packaged with instructions for the DIYer, but the best advice is that only a professional or an advanced DIYer should install siding.

Estimating Siding

To estimate the amount of siding you need, you must calculate the total exterior surface of the house. For a single-story house, double the length of the house and multiply it times 9 (the average height of siding to the soffits). Then double the width of the house and multiply that figure times 9. This will give you the total surface area for a house with a hip roof. If you have gable ends, multiply the length of the house by the height of the gable and divide by 2.

Next, measure the area of the doors and windows. Subtract 21 square feet for each door. (Most exterior doors are 3 feet wide by 7 feet high.) Then measure the width times the height of each window, and add the total area of all windows. Subtract this total door and window area from the total wall area. The result is the total area in square feet of siding needed. Add about 5 percent for waste. This is the amount of siding needed to cover your house.

Wood Siding

Most wood siding is cut from the more durable species. These include cedar, redwood, and cypress, which have natural resistance to rot and insect attack. Other wood siding species include fir, pine, and spruce.

When cutting any type of wood or plywood siding, the material should be coated with water repellent. Keep water repellent and brush handy to coat cut ends or edges on siding.

Wood siding is offered either in vertical or flat grain. This refers to the direction of the log cut in relation to the direction of the wood grain. The most durable wood siding is vertical grain, which is less prone to cracking and warping, and also offers a superior base for paint.

Patterns of wood siding include bevel, in which one edge is thicker than the other; Dolly Varden, in which the bottom edge is rabbeted; and drop siding, which may be plain edged, tongue and groove, or shiplap.

Board-and-batten siding is made up of wide boards (1 × 6 or 1 × 8) with battens of 1 × 2 at the joints. Board-and-batten siding is installed so the boards are vertical on the wall.

Plywood siding is available in drop siding or in sheets that can be installed either vertically or horizontally on the walls. Plywood siding panels may have kerfs or grooves cut at intervals along the length of the panels. Other surface finishes include rough sawn, wood grain, and reverse board-and-batten. Plywood siding panels can also be applied vertically with battens nailed at equal intervals (on joints and either 16 inches or 24 inches on center).

When properly maintained, wood siding offers a traditional appearance, long service, and low maintenance. The downside is that wood siding will require repainting or staining every 5 to 7 years. Keep in mind that quality paint applied strictly according to label directions can extend the interval between paint jobs.

RULE OF THUMB

Siding Finishes

Most wood siding has one smooth side, with the other side rough sawn. For a paintable surface, apply the siding with the smooth side out; for a rustic appearance, apply the siding with the rough side out. Use paint on smooth siding, stain on rough siding.

Metal Siding

Metal siding is either steel or aluminum, both of which are factory-coated with a baked enamel finish for years of freedom from repainting. Both types of metal siding are available in a variety of embossed or wood-grain finishes. Metal sidings were once considered vulnerable to damage from hail or other impact. That objection has been overcome by the use of patterned or wood-grain siding, and by use of backer boards installed behind the siding to stiffen it.

Metal siding edges lock together. At joints or between siding and doors or windows, special channels or strips provide trim edges.

Paint on aluminum siding may oxidize and become dull in time. Use one of the wood-cleaning products designed for cleaning decks to remove the oxidized paint from aluminum siding. Use any quality acrylic latex house paint to paint aluminum siding.

Vinyl Siding

Vinyl siding has become popular because it is inexpensive and maintenance-free. In most areas you will find that, installed, vinyl siding costs about two-thirds the cost of aluminum siding. Some excessive expansion caused problems with early vinyl products, but these problems have been overcome. The nailing tabs on both vinyl and metal siding have slotted nail holes to allow the materials to expand and contract.

Unlike metal siding, which has a factory-applied paint finish, the color in vinyl siding is not a top coat, but is continuous through the product. Vinyl siding never needs repainting. Use a mild solution of TSP to wash down the siding, then hose off with clean water to rinse. Do this at least once a year, in spring or fall.

Wood Shake/Shingle Siding

Walls can be wood-shingled either in a double or single course. In single-course shingle application, the bottom or starter row of shingles is double to provide an even starting point, but the rest of the wall has only a single layer of overlapped shingles.

In double-course shingle application, a first layer of under courses or number 3 (utility-grade) shingles is applied, then number 1 (retrograde) shingles is applied as the finish course. The double-course job results in a deeper shadow line or depth at the bottom edge of each course of shingles.

Decks

A deck (see figure 11.2) can greatly extend your living space during the warm seasons. When properly constructed, a deck will drain quickly after a summer shower, so you can use the deck within minutes after rains cease. The deck is a project that has been successfully completed by thousands of homeowners, and deck plans are offered through lumber dealers by a host of major lumber companies. Some home centers even offer custom-design services to help you plan your deck.

Before beginning your deck, check with your local building department for code requirements and any necessary building permits. Check also to see if any underground utilities, such as electrical or telephone wires, are buried where you will dig deck post holes or piers.

In many areas you will be required to dig post holes down to the frost line, or the depth to which the earth freezes in winter. In the northern tier of states, the frost line is 4 feet deep; the required depth decreases as you move south. In some areas, preformed concrete-block piers are permissible, rather than posts set to frost line. If your deck is fastened to the house by a ledger joist, you

may be able to use the concrete-block piers for support on the outer post supports, because a slight frost heaving is no problem if the deck is fastened directly to the house. If the deck adjoins the house but is free-standing, with support posts at the house side, the posts must be set to frost line, because any frost heaving at the house side of the deck will cause the deck to rise and perhaps block the entry door.

Footings/Posts

If you dig post holes to support the deck, you can choose from a number of ways to install the support posts.

One common method of installing deck support posts is to dig a hole to frost line, and place about 6 inches of gravel in the bottom of the hole. This gravel will help keep water away from the post base, and will help reduce rot of the post. Another choice is to fill the post hole with concrete, and embed a metal post base in the wet concrete. Use fiber tubes (Sonotubes) as forms for the concrete post base. When the concrete has set, use galvanized nails or coated deck screws to fasten the bottom of the wood 4×4 posts to the metal post base.

Railings

Some community building codes permit decks up to 30 inches aboveground to be built without railings. If you have ever stepped off an unseen street curb, you know that stepping unexpectedly off a 30-inch-high deck could be disastrous. We recommend that rails or perimeter seating be built on any deck that is above ground level (less than one step or 10 inches aboveground). Note that various types of perimeter benches or seating can be substituted for railings on decks of any height. Perimeter seating not only serves as a protective barrier at deck edge, but it saves both the cost and the space required for deck furniture.

Figure 11-2 Deck Basics

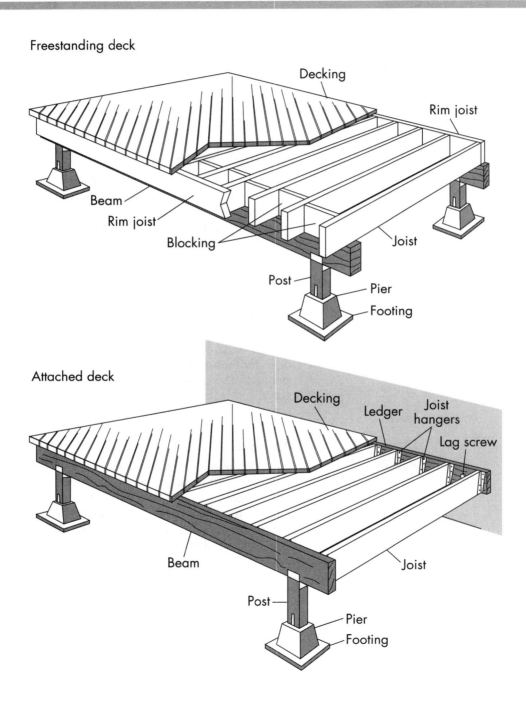

Freestanding deck

Decking

Rim joist

Beam

Rim joist

Blocking

Joist

Post

Pier

Footing

Attached deck

Decking

Ledger

Joist hangers

Lag screw

Beam

Joist

Post

Pier

Footing

Deck Railings

Deck railings should be strong enough to support a person's weight, and should be set 36 to 42 inches above deck level.

Deck Construction

When buying lumber for your deck, choose from the more durable wood species. Woods that are naturally resistant to decay and insect attack include redwood, cedar, and cypress. These wood species are attractive and durable. Only the heartwood of redwood or cedar is decay resistant, so use only heartwood for any lumber that is in ground contact, such as deck posts.

An economical alternative to these species is to use pressure-treated lumber as the support lumber, and build the deck and railings from redwood or cedar as desired.

Deck boards should be either 2 × 4 or 2 × 6; wider deck boards are likely to cup and warp. Rather than choose 2× boards for the deck surface, consider buying ⁵⁄₄-inch-thick deck boards. Deck boards have chamfered edges to reduce splintering along the edges and to permit water to drain quickly from the deck.

It may be difficult to penetrate treated lumber with nails or screws, so consider predrilling holes for nails or screws. Also, some treated lumber may retain a high moisture content when you buy it. Let wet lumber stand in warm weather for a few days to allow it to dry out before you assemble it. This will prevent excessive lumber shrinkage after the deck is built. For example, shrinkage of the deck boards may leave unsightly wide cracks.

Fasteners

To ensure long life for your deck, always build with corrosion-resistant fasteners. The choices are coated deck screws, galvanized nails, lag bolts, stainless steel fasteners, and metal framing connectors.

Fastening structural lumber such as joists by toenailing produces a very weak joint. Instead, use galvanized metal connectors for maximum strength. These metal connectors are available for any application; there are even deck-board tie connectors (see figure 11.3) that permit you to build with no nails or screws in the deck surface boards. Remember that for proper drainage of the deck, there must be a gap between the deck boards. Deck-board tie connectors provide an automatic gauge that provides the proper space between deck boards.

Cantilevered Decks

Some builders choose to extend or cantilever the deck beyond the outboard support posts. Most codes allow the deck to be cantilevered a maximum of 3 feet beyond the support posts; for example, on a deck that is 9 feet wide, 6 feet or two-thirds of the joist length would be between the ledger board on the house side and the outboard posts and beams supporting the deck, with 3 feet cantilevered beyond.

Cantilevered Decks

Because the cantilevered portion of the joists can sag if not properly engineered, homeowner-built decks should have a 3-foot maximum cantilever section.

Figure 11-3 Deck Connector System

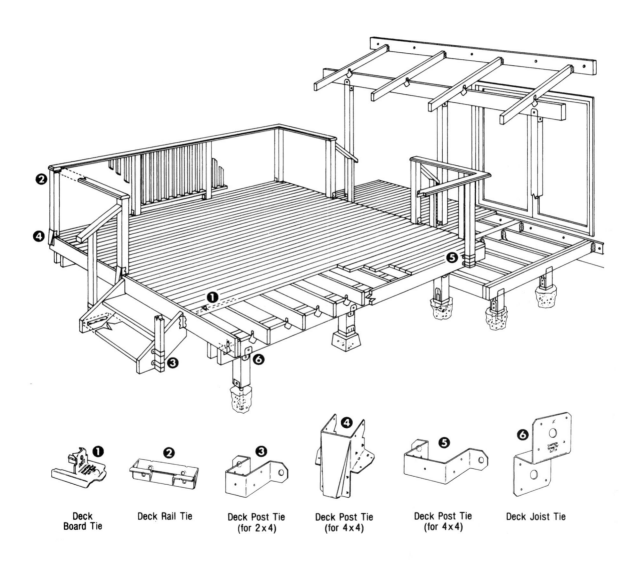

Deck
Board Tie

Deck Rail Tie

Deck Post Tie
(for 2 x 4)

Deck Post Tie
(for 4 x 4)

Deck Post Tie
(for 4 x 4)

Deck Joist Tie

Deck Stairs

Your choice of stairs will depend on the height of the deck aboveground, how much traffic will use the stairs, and the grade or slope of the land where the stairs will be. One approach to easy deck access may be to build a multilevel deck, dropping the deck level as you build to the outer edge.

Fences

Fencing can add security, safety for pets or children, and privacy to your yard. Before building a fence, decide what will be the purpose(s) of the fence, and what style of fence you want—rail, Colonial, picket, wire cyclone, or privacy fence. Then consult your local building department for building code requirements. Most residential fences are 3, 4, or 6 feet high, depending on their purpose.

If your house is newer, a plat map of your lot may be on file at city hall. Or you may be able to locate the metal surveyor's stakes that mark your property boundaries. If you cannot find either a plat map or the surveyor's stakes, have a surveyor do a survey of your lot to determine your exact boundaries. Be aware that any fence built on the neighbor's side of the property line becomes that neighbor's property, and you may be required to take the fence down and move it onto your own property. Your building department can tell you the required setback or required distance from the property line for your fence. Also, most building codes require that if the fence you build does not appear the same on both sides, you must put the best side of the fence outward, toward your neighbor's property.

Fence choices are varied. Wire fences are commonly 4 feet high and are offered in galvanized or plastic-coated steel. Rail fences provide a ranch or rural effect, and were originally intended as a barrier to large animals such as horses. Picket fences are usually 3½ to 4 feet high, and provide a barrier to children and pets. Fencing can be of solid or staggered fence boards, up to 6 feet high. A fence that is 3 or 4 feet high can provide a barrier to wandering pets or children, without blocking the view or the breeze. For complete privacy, a 6-foot-high fence blocks the view of outsiders but also blocks your own view of the area. Figure 11.4 shows a number of fence styles.

To decide what kind of fence to build, visit your home center and inspect the fencing choices there. Home centers offer metal connectors, fence posts, precut pickets, or fence boards, and also have prebuilt privacy fence panels that are 6 feet high and 8 feet long, and are available in a variety of styles. Home centers also sell posts, connectors, rails, wire, and gates for building a cyclone or woven-wire fence.

Fence Posts

Regardless of your choice of fencing, keep in mind that fence posts are subjected to loads, so must be set firmly in concrete. For woven-wire fences the load or stress is produced by stretching the wire between posts. In wood fencing, particularly for solid-wood, 6-foot-high privacy fence panels, the side or wind loading creates high pressure against the sides of the fencing, and transfers these wind loads to the fence

RULE OF THUMB

Fence Post Length

Because of wind and other force loadings, you should always set all fence posts so that one-third their length is in the ground.

Figure 11-4 Fence Styles

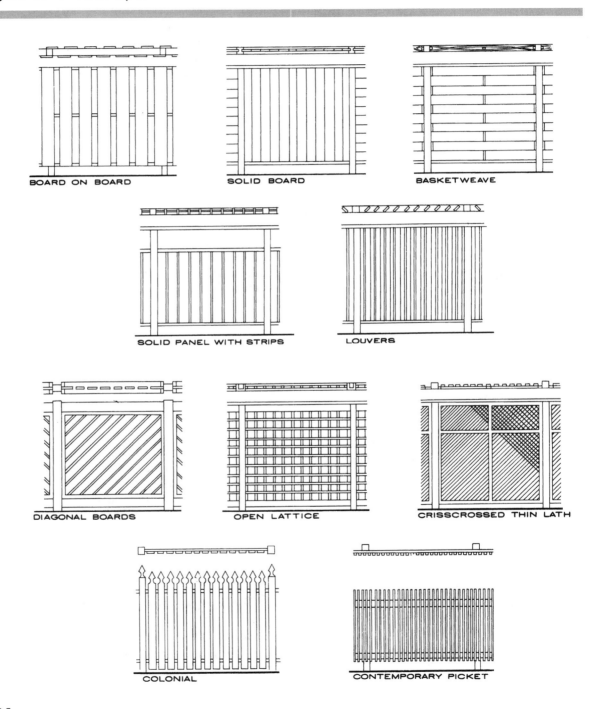

BOARD ON BOARD

SOLID BOARD

BASKETWEAVE

SOLID PANEL WITH STRIPS

LOUVERS

DIAGONAL BOARDS

OPEN LATTICE

CRISSCROSSED THIN LATH

COLONIAL

CONTEMPORARY PICKET

posts. Corner and gate posts are usually braced or otherwise reinforced because of the high stress loads on the fence at these points.

For wood posts, choose from the more durable wood species. Western red cedar, redwood, and pressure-treated lumber are all durable choices for fencing.

To establish the fence line, set a mason's line between the corner posts, then measure and mark your post hole locations. For example, cyclone or wire fencing is built with 10-foot-long top rails, so posts are set 10 feet apart. For most wood fencing you will either use 8-foot-long 2 × 4 rails, or 6-foot-high-by-8-foot-long prebuilt panels, so posts are set 8 feet apart. Plan well to avoid setting the posts at the wrong interval, so posts must be moved. For example, if using 6-by-8-foot fence panels, will the panels be installed so they fall between the posts, or set so they overlap the face of the posts?

Dig the post hole diameter large enough to accept the post plus reinforcing concrete. The amount of reinforcing concrete needed for each post will depend on the type of soil in your area. Heavy clay soil will pack hard, while light sandy loam cannot be packed adequately to help support the posts. Check with your dealer for concrete requirements for supporting fence posts.

When setting wood fence posts, first pour 4 to 6 inches of gravel into the bottom of the post hole. The gravel will let moisture drain away from the post ends and inhibit post rot.

Let the post hole concrete set at least 24 hours (see directions on the concrete-

RULE OF THUMB

Post Hole Concrete

The moisture from the soil around the post hole will be enough to cause the concrete to set. Save work and time by dumping the dry concrete mix into the hole, rather than mix with water. Before dumping in the concrete mix, spray the sides and bottom of the hole with a mist from a hose nozzle to ensure there is enough moisture present to set the concrete.

mix bag) before attaching the fence boards or panels to the fence. Any movement or vibration that causes the post concrete to crack will result in a weak fence post.

Use metal connectors and galvanized fasteners to build the fence. If you wish to provide an entry/exit point through the fence for lawn tractors or other vehicles without building a wide gate, use U-shaped metal hangers to hold the fence rails, but do not nail or screw the fence rails to the metal connectors. Just set the rails loosely into the metal connectors, so that portion of the fence can be removed for access.

When building a wood gate, the most common mistakes are failure to reinforce the wood gate with cross-bracing to prevent sagging, and using too-small hinges and hinge fasteners to secure the gate to the gatepost. A wood gate is heavy; ask your dealer to help you select proper gate hardware.

12 Concrete, Masonry, and Asphalt

Concrete, masonry, and asphalt are among the most durable of home materials. However, their durability does not mean that these materials are maintenance-free. All these products expand and contract as weather and temperatures change. The resulting effects are most damaging in climates where there is a large temperature range. In northern states temperatures may vary from 100 degrees in summer to 30 degrees below zero in winter. The ravages of wear and tear, combined with the assaults of nature, can damage or destroy these materials. This chapter will review rules of thumb for building and maintaining concrete, masonry, and asphalt projects.

Concrete

Most homeowner projects require only small amounts of concrete. For such projects as concrete deck piers, cementing post holes for fences, or replacing a damaged section of a sidewalk, the homeowner should buy bagged pre-mixed dry products, such as Sakrete or Quikrete, at home centers. These products contain a dry mixture of portland cement, sand, and gravel. They are available in quantities (bags) of 10, 40, 60, or 80 pounds, and in a variety of formulations for different applications.

Using the bagged concrete mixes for small projects eliminates the mess of handling the portland cement, sand, and gravel separately. Just add water (follow directions on the bag for water proportion) and mix. A small wheelbarrow can be used for a mixing container, or build a mortar box from plywood. Use a trowel or a garden hoe to stir the mix.

RULE OF THUMB

Mixing Cement

When mixing cement, always pour the dry mix into the container, then add the water to the dry mix. Add water slowly to avoid having too much water and a too thin mix.

For larger concrete projects consider buying the cement premixed. If you buy the cement and have it delivered by a ready-mix truck, you will save the work of mixing and moving the cement (the driver will pour the cement via a long chute, placing the mix where needed), plus you will avoid the problem of disposal of leftover ingredients such as sand and gravel.

Another advantage of buying ready-mix concrete is that the plant will proportion the cement-sand-gravel-water ingredients properly for the job and conditions you specify. For example, when the pour will be done in very hot weather, the mixing plant will add more water to the mix, so it is a bit wetter and will not get too stiff to pour and finish.

If you decide to mix the cement yourself, you can rent a cement mixer from tool rental outlets. You can tow the mixer home via a trailer hitch. For home mixing you must buy the cement in 80- or 94-pound bags, and buy the sand and gravel in bulk. Again, to avoid getting a watery mix, pour the dry products into the mixer and add water sparingly as the mixer turns. Keep a garden hose with a nozzle ready at the mixer. Use the hose to add the water to the mix, and to clean the mixer between batches. Keep in mind that the amount of water needed can vary because of the water content of the sand.

Estimating Needs

Concrete is sold by the cubic yard. A cubic yard is 36 inches long by 36 inches wide by 36 inches high. To estimate the amount of concrete needed for a slab, decide how thick your slab will be. For example, slabs for patios, driveways, and sidewalks usually are formed with 2 × 4s, so the slab is about 4 inches thick. Because the surface area is 1 square yard (36 × 36 inches), just divide the yard-square cube height (36 inches) by the thickness of the planned slab. In this case, 36 inches divided by a 4-inch slab thickness is 9, so a cubic yard of concrete will cover 9 square yards to the 4-inch slab depth. If you are pouring a 6-inch-thick slab, divide 36 inches by 6 inches to get 6 square yards of coverage per cubic yard.

Forms

To hold and shape concrete, you must build forms. If you are preparing concrete piers to support a deck, buy rigid fiber tubing (trade names are QUICK-TUBEs or Sono-tubes) for forms. Just dig the pier hole to the proper depth—to the frost line. (Your local dealer or building inspector can tell you the frost depth—i.e., the depth to which the soil will freeze—for your area.) In southern climes this may be only 1 foot; in northern states the frost line is 4 feet deep. Pour 6

RULE OF THUMB

Proportions

When mixing concrete, remember the 1-2-3 rule. This rule means that for most jobs you add one part cement, two parts sand, and three parts gravel, with enough water to bring the mix to the desired stiffness or slump rate. How to measure these products? Professionals make it easy: use the shovel, and simply add and mix one shovelful of cement, two shovelfuls of sand, and three shovelfuls of gravel, or multiples of these amounts depending on the amount of concrete needed.

inches of gravel into the bottom of the hole. Cut the form tubes to the correct length, meaning the depth of the hole plus the height you wish the piers to sit above grade, and place them in the holes. Be sure the tops of the forms are level with one another, so that the support beams and joists will sit level on the piers. To check the level of the tops of the tube forms, attach a mason's line running to the top edge of each form. Then use a line level (a small lightweight level that can be hooked to the mason's line) to check for level from the top of one tube to the top of the others. Adjust the height of the form tubes by adding more gravel to the bottom of the hole. When the tops of the form tubes are level with one another, pour the concrete into the tubes.

For forming slabs, use 2 × 4s as forms. Use mason's string on stakes to mark the boundaries of the slab. Then position the

RULE OF THUMB

Checking the Square of Forms

To check the square of forms, measure diagonally from opposite corners (see figure 12.1). If these dimensions are the same, the form is square.

2 × 4s and drive steel or wood support stakes along the outside edges of the 2 × 4s.

Use plywood for forming steps or porches, where larger form materials are required. Keep in mind that wet concrete is very heavy, and form supports must be sufficiently strong to hold the weight of the concrete without bending or sagging.

Figure 12-1 Checking the Square of Concrete Forms

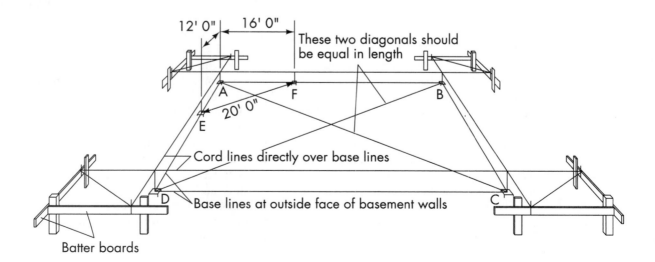

Use double-headed form nails for easy disassembly of the forms and stakes or braces. You must seal the bare form boards so that the concrete will not stick to them and leave an uneven surface when forms are pulled away. Professionals once used light motor oil or form oil to paint the form boards; a more environmentally friendly method would be to coat the side of the forms that will be next to the concrete with a water repellent/sealer such as that used on wood decks. Let the sealer dry before placing the form boards.

On concrete slabs, do not be in a hurry to strip the form boards away, because the forms will help protect the edges of the slab from damage until the concrete has cured.

For formed surfaces where you desire a rubbed finish (rub the still-wet concrete on the exposed concrete with a sponge float), such as the sides or edges of steps or porches, you must remove the forms before the concrete has completely set. Depending on working temperatures, concrete will be set enough to strip forms after about 8 hours. By doing this you can rub or float-finish the concrete with a wet sponge float, and fill any voids caused by the forms on the sides of steps or porches.

Curing Concrete

Concrete should not dry, but should cure in the presence of water. Instructions for curing concrete once advised to cover the concrete project with hay or wet burlap to slow water loss and improve the curing process. Today, the best advice is to cover the freshly laid slab with plastic sheets. Or apply a concrete sealer to the surface as soon as the concrete is set hard but not dry, which will usually be late in the day after a morning pour. Use a pump-type garden sprayer to apply water-thin concrete sealers to concrete surfaces.

Sealing Concrete

Although concrete is very durable, it is wise to use a concrete sealer on all concrete slab surfaces. Sealing concrete slabs will prevent dusting of the slab surface, thus reducing housecleaning chores caused by tracking of concrete dust; will protect the driveway and garage floor from freezing water and from road salt picked up on icy roads; and will make it easier to clean up any spills or stains on the concrete. Look for a quality concrete sealer at your dealer, or at janitorial supply stores.

Repairs

Cracks

It is essential to repair cracks that develop in any concrete surface. If cracks are left open, water will enter and erode the sand/gravel base under the slab. Also, in cold weather the water will freeze and expand, further damaging the slab.

The common advice for repairing cracks in concrete is to use a chisel to undercut the crack, clean the void to remove dirt and debris (a Shop-Vac is handy for this purpose), and then to mix concrete patcher and use a trowel to fill and smooth the cracks. For most cracks, those up to $\frac{3}{8}$ inch wide, the simple solution is to clean the crack and remove any loose or damaged portion of the concrete. For very deep cracks, use clean sand to fill the void to within $\frac{1}{2}$ inch of the top. Then use a concrete patcher in a caulk tube to patch the crack. Or choose one of the epoxy concrete patching kits. Because concrete will expand and contract with temperature changes, no crack repair can be considered permanent, so patching cracks in concrete is a maintenance job that must be repeated each year.

Stairs, Steps

The most common repair problem with concrete steps or stairs is broken step corners. The damage may be caused by impact, for example, striking the edge of the step when moving a heavy object such as a home appliance. This is a difficult repair for the amateur, because it is difficult to form a new corner on the step and to get a good bond between the damaged concrete and the repair material. One possible DIY solution is to clean the old concrete surface and apply a liberal coat of concrete bonding liquid (available at home centers) to the concrete surface. Then bore holes in the old concrete and insert steel reinforcing rod (called rerod) or attach wire lath to support the concrete patch material. Keep in mind that it is difficult to get a patch that will not break again, and the best advice is to hire a professional to do the repair.

Sidewalks

Because a hole in a sidewalk is on a horizontal surface, gravity is working for you so that patch materials are easier to place. For holes in sidewalks, remove any loose concrete or dirt, and use a Shop-Vac to clean away the debris. Use a paintbrush to apply a concrete bonding liquid to surfaces of the hole in the old concrete. Then mix the concrete patcher according to directions, and use a flat trowel to place and smooth the concrete. If you are using a prepared concrete patching kit, follow the package instructions for your particular product.

Small Holes in Concrete

● Clean out the hole and wet it before applying the concrete.

● Mix one part portland cement with two-and-one-half parts of fine sand, adding enough water so that the mix will form a ball when squeezed in your hand.

● Use a pointing trowel to place and smooth the concrete patch.

When a sidewalk is poured and finished, the worker uses a grooving tool to cut expansion joints at intervals along the slab. These joints permit the sidewalks to crack in a straight line when the slab expands in hot weather, rather than break in an irregular or crooked line. If one section of sidewalk is badly broken or cracked, you can break out the entire section and replace it.

Use a mason's chisel to break the portion of concrete between two expansion cracks or grooves. Remove the pieces of broken concrete. If the sidewalk section has heaved due to frost or to a tree root growing beneath the walk, dig out the earth or the tree root that caused the heaving and cracking. If there is little gravel base under the old walk, dig a trench as wide as the walk and level the bottom of the trench.

Set 2 × 4 forms on both sides of the existing walk to mark the edges of the new slab. Then add 4 inches of gravel to the bottom of the trench. Compact the gravel by walking on it, or use a concrete block to compact the gravel. Tie a rope to the concrete block and alternately raise the block and let it fall against the gravel. Also, soaking the gravel or earth with a water hose will help compact the gravel fill.

When the form area is prepared, fill it with concrete. Cut a 2 × 4 that is a foot longer than the width of the walk. Use the 2 × 4 as a screed to level the concrete. Set one end of the 2 × 4 over each of the parallel form boards, and pull the 2 × 4 along the top of the form boards to smooth and level (screed) the concrete. Let the concrete set a bit, until there are no standing water puddles on top of the slab, then trowel the surface of the concrete smooth. Use a hand tool called an edging tool to break or round over the edges or corners of the slab. When the concrete has set enough so that it will hold its shape, use a broom to broom the surface; that is, sweep the broom across the slab so that small swirls or grooves are left in the surface. This broomed finish will provide better traction on wet concrete than a trowel-smooth finish. Cover the slab with plastic sheeting to hold in the moisture for curing and to keep dirt off the new slab.

Masonry

Brick and Stone

Any brick or stone project that is unevenly laid will present an amateurish effect and will detract from your home's appearance. Laying brick or stone with mortar is a job that should be left to professionals, because maintaining proper plumb, level, and mortar joint spacing is a difficult job for the novice.

Some brick walks or patios are laid in sand rather than mortar, and this type of brick construction can be done by the homeowner. Use graph paper to draw a plan of the project, allowing about a 3/16-inch gap between the bricks. Set 2 × 4 forms as boundary guides for the project. Set the bricks within the form so they are evenly spaced. When the bricks are set in the desired pattern, sweep fine sand over the surface, filling the joints between the bricks. Use a fine mist of water from a garden hose nozzle to pack the sand, then repeat the sweeping process to ensure the cracks are filled level to the brick tops with sand.

For the reasons mentioned above, you should leave to the professionals jobs such as tuck-pointing the mortar joints between courses of brick or stone. The mortar joints will darken in time, and the DIYer does not have the skills necessary to match the old mortar color. New mortar mix placed in cracked joints will usually be much lighter or whiter than the existing mortar, and the tuck-pointing job will give the house a striped or zebra effect.

Brick or stone will gray and become dirty with time. You can give the brick or stone project a like-new appearance by pressure-washing it. A good procedure is to first wet down the entire wall, and let the water set a few minutes so it can lift and loosen the dirt and grime. Then, using the pressure sprayer, apply a masonry cleaning product to the surface as per label directions. (Professional-quality masonry cleaners are available at janitorial supply stores.) When the cleaner has been given time to work, use the pressure washer to rinse away the grime and cleaner residue. You can rent a pressure washer at most tool rental outlets.

Cleaning Mortar from Brick or Stone

Professionals use muriatic acid wash to clean fresh mortar from a newly laid surface, but the acid is useless for removing mortar once the mortar is old or dried. Never use muriatic acid on porous concrete block, because the acid may soak into the block and cause damage.

Concrete Block

Large-scale block projects such as basements are beyond the scope of the do-it-yourselfer. However, small projects such as a block partition in a basement or a backyard greenhouse can be done by the handyman.

Concrete blocks are usually 15½ inches long, 8 inches high, and 3½ to 12 inches wide. Houses built in the fifties may have basement walls laid with 8-inch block, but most block basements today are made of 12-inch block.

Mortar not only bonds blocks or bricks together, but also helps make the wall watertight and permits the workman to lay blocks in level courses, even though the measurements of individual blocks may vary slightly. Mortar mix is available from most home centers, premixed in 40-pound bags. Although a mortar of portland cement and sand can be used for laying blocks, a better mortar will also contain lime. Lime added to the mix makes the mortar cling better to tools and to the block or brick, and is easier to handle and spread than cement/sand alone. Cement/lime mortar does not sag or drop from the joints as the blocks are laid, will prevent the blocks from shifting position once laid, and will cling to vertical surfaces such as the end joints between the blocks.

When building a block partition in a basement—for example, building a block shower stall—use 3½-inch-wide blocks. For a lightweight partition wall, the blocks can be laid directly on the concrete floor; no footings are needed. Lay out the job on graph paper first, to avoid having to cut block and to ensure a workmanlike job. Measure the distance from the concrete basement floor to the basement ceiling, and divide this distance in inches by the height or thickness of the block. Allow ⅜ inch between each block course for the mortar joint.

Mark the outside perimeter of the wall(s) on the concrete floor. Use a mortar mix that can be purchased at a home center. Mix the mortar according to label directions, keeping the mix stiff enough so that it can be handled on the trowel. Using the trowel, lay a bead of mortar along both inside and outside edges where the block will sit on the floor. To avoid rapid drying and weak mortar joints, place only enough mortar to lay each block. Place the block atop the mortar beads and press it firmly in place, leaving a mortar joint that is ⅜ inch

thick, and use the trowel to remove excess mortar from the edges of the block.

Next, lay the mortar for the next block on the concrete floor, then lay a bead of mortar on the edges of the end of the first block to join the block ends together. Use a 2-foot level to check for level and plumb at each block. Keep the job clean as the wall rises, being careful not to smear mortar on the face of the block. Do not use an acid wash to clean concrete blocks. It is difficult to remove spilled mortar from the block faces. If you drop a blob of mortar on the block face, let it set until it is firm before using the trowel edge to clean it away.

Proceed in the same way to lay the walls up. Never move or shift any block after the block is positioned, leveled, and plumbed. Moving the block can break the bond between the block and the mortar, resulting in a weak wall and failing joints. Be careful to maintain mortar joints that are uniformly ⅜ inch thick. At wall corners the blocks from the two walls must overlap with alternate courses. Use end blocks with square edges at the sides of any door.

When building outdoor block projects such as a backyard greenhouse, it is necessary to pour a concrete footing to support the weight of the blocks. Dig a trench to the frost line—as noted before, the frost line in the United States varies from a depth of 1 foot in the south and far west to 4 feet in the northern tier of states. Your local building inspector can advise you of the depth of the frost line in your area.

Dig the footing trench so the bottom is level. The footing should be poured on undisturbed soil. Footings poured on loose-fill dirt may crack and fail when the loose soil settles.

The rule of thumb is that a concrete footing for block should be as thick as the wall is thick, and twice as wide as the wall is wide. For example, a wall built of 8-inch block should be built over a footing that is

8 inches thick and 16 inches wide. For large projects such as a house, the building codes require that to further strengthen the footing, steel reinforcing rod, called rerod, must be placed in the concrete footing. Check with the building inspector to learn if rerod is required for your project.

Use wood form boards to form the project footings. Pour concrete in the forms and use a 2 × 4 longer than the distance between the forms to screed and level the top of the footings. Let the concrete footings cure before beginning to lay the block foundation.

Asphalt

Because asphalt requires heavy and expensive equipment for placing, rolling, and compacting the material, laying asphalt is not a do-it-yourself project. However, you can do most asphalt maintenance and repair jobs yourself.

Pothole Repair

Before beginning asphalt pothole repair, clean away any debris from the hole. Use a mason's chisel to chisel away cracked asphalt along the edge of the hole. The pothole should be clean and dry before repair is begun.

Asphalt cold patcher is available in 40-pound bags at home center stores. The asphalt often hardens in the bag so it is difficult to pour and smooth. To soften the asphalt patcher, use a heat gun such as those sold for paint removal. Play the hot air stream back and forth over the mix to soften it. Then use a trowel to place and smooth the patcher.

Filling Cracks

For filling minor cracks in asphalt, up to $3/8$ inch wide, use the patcher in a caulk tube, available from home centers. Use a shop vac to clean loose material from the crack, then use the caulk gun to fill the crack. Use a plastic spoon or spatula to smooth the patch material.

To fill deep asphalt cracks, use clean sand to fill the cracks within an inch of the surface. Then fill the cracks with pourable crack patcher, available at home centers. Pour slowly, until the crack has been completely filled.

Sealing Asphalt

New asphalt does not require a seal coat. Indeed, there is an argument whether asphalt sealer is ever needed or is merely used as a cosmetic touch to blacken and renew the asphalt surface. But as the asphalt slab ages, the hot sun bakes the oils out of the slab, and cracks develop from wear and weathering of the slab. Especially in cold climates, water can enter the cracks and freeze, expand, and break the slab. To protect the slab from water entry, the asphalt should then be sealed. The most common asphalt sealer is an emulsion or latex sealer that is sold in 5-gallon pails at home centers.

To Apply Emulsion Sealer

- Clean the driveway with a commercial driveway cleaner product or a non-sudsing detergent such as TSP (trisodium phosphate, available at paint stores), rinse with clear water, and let the slab dry.

- Mix the sealer to put the solids into suspension, and pour a gallon or so onto the slab.

- Spread the sealer with an inexpensive broom/squeegee tool, sold at home centers. Proceed in this manner until the entire slab is sealed.

One drawback of ordinary emulsion or latex sealers is that they leave a surface film, just as paint does, on the asphalt surface. After repeated coats of sealer, this surface film will begin to develop random cracks, called "alligatoring." It is impossible to remove the alligatored surface film. Therefore, a better sealer choice is to use an all-petroleum asphalt sealer. These petroleum-base sealers can be purchased from many asphalt contractors, or look for them at your janitorial supply shop (see the Yellow Pages under Janitorial Equipment & Supplies).

Hot-mop petroleum sealers are available at asphalt mixing plants. There you can purchase the same hot sealer that contractors spray on road surfaces or commercial parking lots. For obvious reasons, hauling hot mix can be dangerous. Use a metal garbage can and lid, not a plastic one, to haul the hot sealer home. Use a bungee cord to secure the lid and tie the garbage can so it cannot tip over. To avoid accidents, do not try to haul hot mix in a car. If you do not have a pickup truck or trailer, pay the professionals to seal the driveway with hot mix.

Index

About the Authors

Gene and Katie Hamilton are creators of HouseNet, an online service about home improvement information on America Online (keyword: housenet) and on the Internet (www.housenet.com).

Their weekly newspaper column, "Do It Yourself or Not?" is syndicated by the Los Angeles Times Syndicate.

They live on the eastern shore of Maryland.